NAVY'S NEEDS IN SPACE FOR PROVIDING FUTURE CAPABILITIES

Committee on the Navy's Needs in Space for Providing Future Capabilities
Naval Studies Board
Division on Engineering and Physical Sciences

NATIONAL RESEARCH COUNCIL
OF THE NATIONAL ACADEMIES

THE NATIONAL ACADEMIES PRESS
Washington, D.C.
www.nap.edu

THE NATIONAL ACADEMIES PRESS 500 Fifth Street, N.W. Washington, DC 20001

NOTICE: The project that is the subject of this report was approved by the Governing Board of the National Research Council, whose members are drawn from the councils of the National Academy of Sciences, the National Academy of Engineering, and the Institute of Medicine. The members of the committee responsible for the report were chosen for their special competences and with regard for appropriate balance.

This study was supported by Contract No. N00014-00-G-0230, DO #16, between the National Academy of Sciences and the Department of the Navy. Any opinions, findings, conclusions, or recommendations expressed in this publication are those of the author(s) and do not necessarily reflect the views of the organizations or agencies that provided support for the project.

International Standard Book Number 0-309-09677-4

Additional copies of this report are available from:

Naval Studies Board, National Research Council, The Keck Center of the National Academies, 500 Fifth Street, N.W., Room WS904, Washington, DC 20001; and

The National Academies Press, 500 Fifth Street, N.W., Lockbox 285, Washington, DC 20055; (800) 624-6242 or (202) 334-3313 (in the Washington metropolitan area); Internet, http://www.nap.edu.

Printed in the United States of America

THE NATIONAL ACADEMIES

Advisers to the Nation on Science, Engineering, and Medicine

The **National Academy of Sciences** is a private, nonprofit, self-perpetuating society of distinguished scholars engaged in scientific and engineering research, dedicated to the furtherance of science and technology and to their use for the general welfare. Upon the authority of the charter granted to it by the Congress in 1863, the Academy has a mandate that requires it to advise the federal government on scientific and technical matters. Dr. Ralph J. Cicerone is president of the National Academy of Sciences.

The **National Academy of Engineering** was established in 1964, under the charter of the National Academy of Sciences, as a parallel organization of outstanding engineers. It is autonomous in its administration and in the selection of its members, sharing with the National Academy of Sciences the responsibility for advising the federal government. The National Academy of Engineering also sponsors engineering programs aimed at meeting national needs, encourages education and research, and recognizes the superior achievements of engineers. Dr. Wm. A. Wulf is president of the National Academy of Engineering.

The **Institute of Medicine** was established in 1970 by the National Academy of Sciences to secure the services of eminent members of appropriate professions in the examination of policy matters pertaining to the health of the public. The Institute acts under the responsibility given to the National Academy of Sciences by its congressional charter to be an adviser to the federal government and, upon its own initiative, to identify issues of medical care, research, and education. Dr. Harvey V. Fineberg is president of the Institute of Medicine.

The **National Research Council** was organized by the National Academy of Sciences in 1916 to associate the broad community of science and technology with the Academy's purposes of furthering knowledge and advising the federal government. Functioning in accordance with general policies determined by the Academy, the Council has become the principal operating agency of both the National Academy of Sciences and the National Academy of Engineering in providing services to the government, the public, and the scientific and engineering communities. The Council is administered jointly by both Academies and the Institute of Medicine. Dr. Ralph J. Cicerone and Dr. Wm. A. Wulf are chair and vice chair, respectively, of the National Research Council.

www.national-academies.org

COMMITTEE ON THE NAVY'S NEEDS IN SPACE FOR PROVIDING FUTURE CAPABILITIES

ANTONIO L. ELIAS, Orbital Sciences Corporation, *Co-Chair*
WILLIAM D. SMITH, Fayetteville, Pennsylvania, *Co-Chair*
ALAN BERMAN, Applied Research Laboratory, Pennsylvania State University
E. ANN BERMAN, Tri-Space, Inc.
THOMAS C. BETTERTON, Naval Postgraduate School
CHARLES F. BOLDEN, JR., Houston, Texas
JOHN F. EGAN, Nashua, New Hampshire
BRIG "CHIP" ELLIOTT, BBN Technologies
RICHARD FLEETER, AeroAstro, Inc.
LEE M. HAMMARSTROM, Applied Research Laboratory, Pennsylvania State University
DONALD G. HARD, Springfield, Virginia
ROBERT E. LINDBERG, National Institute of Aerospace
GEORGE O. NOSSAMAN, BAE Systems
C. KUMAR N. PATEL, Pranalytica, Inc.
GENE H. PORTER, Nashua, New Hampshire
JOSEPH B. REAGAN, Saratoga, California
DWIGHT C. STREIT, Northrop Grumman Space Technology
H. GREGORY TORNATORE, Ellicott City, Maryland
DAVID A. WHELAN, The Boeing Company
DELL P. WILLIAMS III, Moss Beach, California

Staff

RONALD D. TAYLOR, Director (on leave as of July 12, 2003)
CHARLES F. DRAPER, Acting Director (as of July 12, 2003)
MICHAEL L. WILSON, Study Director (through August 27, 2004)
MARY G. GORDON, Information Officer
SUSAN G. CAMPBELL, Administrative Coordinator
IAN M. CAMERON, Research Associate
SIDNEY G. REED, JR., Consultant

Staff

RONALD D. TAYLOR, Director (on leave as of July 12, 2003)
CHARLES F. DRAPER, Acting Director (as of July 12, 2003)
ARUL MOZHI, Senior Program Officer
MARY G. GORDON, Information Officer
SUSAN G. CAMPBELL, Administrative Coordinator
IAN M. CAMERON, Research Associate

Preface

The growing importance of space to the nation's security and economic well-being demands that the United States, through its defense and intelligence establishments, actively pursue policies to maintain and develop capabilities which ensure that the nation's civil, commercial, defense, and intelligence establishments can operate successfully in space. Furthermore, assured access to space and space assets is necessary as the nation's armed forces move toward increased reliance on information and networking technology to integrate decision makers, sensors, forces, and weapons into a highly adaptive and comprehensive system, and thus to increase mission effectiveness.

In 2001, the Commission to Assess the U.S. National Security Space Management and Organization amplified this theme of U.S. dependence on space and recommended a major restructuring of National Security Space (NSS) management and organization.[1] Specifically, the Space Commission recommended, and shortly thereafter the Secretary of Defense directed, that the Secretary of the Air Force serve as the Department of Defense (DOD) Executive Agent for all NSS programs and that the National Reconnaissance Office (NRO) be realigned accordingly.[2] The responsibilities of the DOD Executive Agent for Space are currently delegated to the Under Secretary of the Air Force, who also serves as the director of the NRO. As a result, the Air Force has primary responsibility for

---

[1]Commission to Assess United States National Security Space Management and Organization. 2001. *Report of the Commission to Assess United States National Security Space Management and Organization*, Washington, D.C., January 11.

[2]Paul Wolfowitz, Deputy Secretary of Defense. 2003. "DOD Executive Agent for Space," DOD Directive 5101.2, Department of Defense, Washington, D.C., June 3.

developing, acquiring, and fielding NSS systems that meet the operational needs of the DOD, and the NRO has primary responsibility for developing, acquiring, and fielding NSS systems for the wider national security effort as well as for providing support to the DOD.

The Department of the Navy is a major user of space capabilities. Historically, the Navy filled a significant role in developing a broad range of space capabilities: navigation, communications, environmental, and other systems. These capabilities are now provided predominantly by the DOD, the Air Force, and the National Oceanic and Atmospheric Administration (NOAA). Recent military operations in Bosnia, Kosovo, Afghanistan, and Iraq demonstrated again the need for naval forces to rely on national and commercial space capabilities (which played an essential role in joint operations).

Subsequent to issuance of the 2001 Space Commission report, the Under Secretary of the Navy commissioned the Panel to Review Naval Space "to make a careful assessment of the Department of the Navy space policy and strategy to ensure that the maximum amount of operational space support is provided to naval warfighters."[3] In March 2002, the panel recommended that the Department of the Navy provide senior oversight, management, and participation in NSS organizations. In addition, the panel recommended that the Department of the Navy increase its technology investments in space, revitalize its space systems requirements process, and refocus the management of its naval space cadre. The present study serves, in part, as an extension of the work of the Panel to Review Naval Space. At issue are the Department of the Navy's needs in space for providing future capabilities, taking into account its role in influencing the operational, technical, programmatic, and budgetary aspects of NSS programs.

TERMS OF REFERENCE

At the request of the Chief of Naval Operations, the National Research Council, under the auspices of the Naval Studies Board, conducted a study to examine the Department of the Navy's needs in space for providing future operational and technical capabilities. Taking into account that joint operations will evolve to be more dependent on future naval capabilities, the study addressed the opportunities offered by and the implications of space, as well as addressing the following:

- Review future naval operational concepts (Sea Strike, Sea Shield, Sea Basing, and FORCEnet) and plans for naval (and joint) force use of space, and identify corresponding specific space and space-related needs;

[3]Panel to Review Naval Space. 2002. *Report of the Panel to Review Naval Space: Assured Space Capabilities for Critical Mission Support*, Center for Naval Analyses, Alexandria, Va., March 19.

- Evaluate corresponding naval space-related requirements as a means to employ those future naval operational concepts and plans, and compare the relation of space-related requirements to other requirements that compete for Department of the Navy funding;
- Examine the results of past and ongoing experiments designed to address naval unique space needs, and recommend future opportunities in experimentation for enhancing operational and technical system space-related capabilities; and
- Assess the Department of the Navy's space technical expertise and space science and technology base, and identify research priorities in space for supporting future naval operational concepts.

THE COMMITTEE'S APPROACH

The committee's approach for examining the Department of the Navy's needs in space for providing future capabilities is rooted in the first item in the terms of reference listed above. Specifically, by integrating the Navy and the Marine Corps visions—Sea Power 21[4] and Marine Corps Strategy 21,[5] respectively—the Naval Operating Concept for Joint Operations, known informally as the NOC, describes how the Naval Services "will organize, deploy, employ, and sustain forces to conduct operations guided by the interrelated and complementary concepts of *Sea Strike*, *Sea Shield*, and *Sea Basing* integrated with the family of Marine Corps concepts, *Expeditionary Maneuver Warfare*, *Operational Maneuver From the Sea*, and *Ship-to-Objective Maneuver*; all of this will be enabled by *FORCEnet*."[6] As a result of this integration, the NOC's description of the Navy's Sea Strike concept, for example, takes into consideration Marine Corps concepts such as Operational Maneuver from the Sea and Ship-to-Objective Maneuver. In addition, the Department of the Navy's capstone vision, Naval Power 21, articulates the naval transformational vision and integrates Sea Power 21 and Marine Corps Strategy 21 for the sea-air-land-and-space domain.[7] And the recently released *Marine Corps Concepts and Programs 2004* states that

---

[4]ADM Vern Clark, USN, Chief of Naval Operations. 2002. "Sea Power 21," *U.S. Naval Institute Proceedings*, Vol. 128, No. 10, pp. 32-41.

[5]Gen James L. Jones, USMC, Commandant of the Marine Corps. 1999. *Marine Corps Strategy 21*, Department of the Navy, Washington, D.C., July.

[6]ADM Vern Clark, USN, Chief of Naval Operations; and Gen Michael W. Hagee, USMC, Commandant of the Marine Corps. 2003. *Naval Operating Concept for Joint Operations*, Department of the Navy, Washington, D.C., September 22, p. 3.

[7]Gordon England, Secretary of the Navy; ADM Vern Clark, USN, Chief of Naval Operations; and Gen James L. Jones, USMC, Commandant of the Marine Corps. 2002. *Naval Power 21 . . . A Naval Vision*, Department of the Navy, Washington, D.C., October.

"seabasing effectively integrates the transformational thrust of Marine Corps Strategy 21 and the Navy's Sea Power 21 visions."[8] Accordingly, for the purposes of this report, the committee elected to structure its examination of the Department of the Navy's needs in space in terms of the Navy's Sea Power 21 vision (i.e., those needs arising from Sea Strike, Sea Shield, Sea Basing, and FORCEnet).

While the committee had access to a broad range of information on Navy and Navy-related space technology programs and systems, it was unable to fully address the charge in the second of the tasks listed above: to "compare the relation of space-related requirements to other requirements that compete for Department of the Navy funding." Thus, the committee made recommendations consistent with establishing priorities among space-related requirements but did not seek to address priorities of space- versus non-space-related requirements.

The Committee on the Navy's Needs in Space for Providing Future Capabilities first convened in June 2003 and held additional meetings over a period of 6 months, both to gather input from the relevant communities and to discuss the committee's findings.[9] Agendas for these meetings are provided in Appendix F.

The months between the committee's last meeting and the publication of the report were spent preparing the draft manuscript, gathering additional information, reviewing and responding to the external review comments, editing the report, and conducting the required security review necessary to produce an unclassified report.

[8]Headquarters, U.S. Marine Corps. 2004. *Marine Corps Concepts and Programs 2004*, Quantico, Va., p. 6.

[9]During the entire course of its study, the committee held meetings in which it received (and discussed) classified materials. Accordingly, the information contained in this report has been restricted in order to produce an unclassified report.

Acknowledgment of Reviewers

This report has been reviewed in draft form by individuals chosen for their diverse perspectives and technical expertise, in accordance with procedures approved by the National Research Council's (NRC's) Report Review Committee. The purpose of this independent review is to provide candid and critical comments that will assist the institution in making its published report as sound as possible and to ensure that the report meets institutional standards for objectivity, evidence, and responsiveness to the study charge. The review comments and draft manuscript remain confidential to protect the integrity of the deliberative process. We wish to thank the following individuals for their review of this report:

Timothy P. Coffey, National Defense University,
Donald L. Cromer, Lt Gen, USAF (retired), Lompoc, California,
Paul A. Fratarangelo, MajGen, USMC (retired), St. Pete Beach, Florida,
Robert J. Hermann, Global Technology Partners, LLC,
William E. Howard, McLean, Virginia,
Alfred U. MacRae, Berkeley Heights, New Jersey,
John C. Mather, NASA Goddard Space Flight Center,
Marc Y.E. Pelaez, RADM, USN (retired), Williamsburg, Virginia, and
John F. Vesecky, University of California, Santa Cruz.

Although the reviewers listed above provided many constructive comments and suggestions, they were not asked to endorse the conclusions or recommendations, nor did they see the final draft of the report before its release. The review of this report was overseen by Lee Hunt, Alexandria, Virginia. Appointed by the

National Research Council, he was responsible for making certain that an independent examination of this report was carried out in accordance with institutional procedures and that all review comments were carefully considered. Responsibility for the final content of this report rests entirely with the authoring committee and the institution.

Contents

APPENDIXES

Executive Summary

TODAY'S FRAMEWORK IN PERSPECTIVE

Today U.S. naval forces are dependent on space systems in many ways—for communications; command and control; intelligence, surveillance, and reconnaissance (ISR); navigation; and meteorology and oceanography (METOC)—and they will be even more so in the future. In particular, the Department of the Navy has witnessed significant changes in both the national security environment[1] and the Department of Defense (DOD) posture toward the support and development of space technologies. For example, the homeland defense mission recently established by the Chief of Naval Operations (CNO) tasks the Navy (in concert with the U.S. Northern Command, the Coast Guard, and other federal agencies) to develop new systems and procedures to protect the maritime approaches to the United States.[2] While the Navy's specific roles and responsibilities in support of this mission remain undefined, warranting clarification, they will likely entail new and greater reliance on space—for example, to supply near-continuous, open-ocean surveillance of all surface craft.

To refocus Department of the Navy attention for purposes of these new missions, the Naval Services have recently promulgated a series of capstone vision documents. These include the Department of the Navy capstone vision, Naval Power 21;[3] its succeeding operational extension, the Naval Operating Con-

---

[1]Including current efforts at nation building and counterterrorism.

[2]ADM Vern Clark, USN, Chief of Naval Operations. 2004. "CNO Guidance for 2004," Department of the Navy, Washington, D.C., February.

[3]Gordon England, Secretary of the Navy; ADM Vern Clark, USN, Chief of Naval Operations; and Gen James L. Jones, USMC, Commandant of the Marine Corps. 2002. *Naval Power 21 . . . A Naval Vision*, Department of the Navy, Washington, D.C., October.

cept for Joint Operations;[4] and the preceding Navy and Marine Corps vision documents, Sea Power 21[5] and Marine Corps Strategy 21,[6] respectively. In general, these documents show a progression of visions that has integrated most of the Service-specific concepts into a broader structure founded on the four concepts presented in Sea Power 21—Sea Strike, Sea Shield, Sea Basing, and FORCEnet.

Using Sea Power 21 as an operational framework and taking into account new missions, such as the provision of a joint task force afloat headquarters or ocean surveillance for homeland security, the Committee on the Navy's Needs in Space for Providing Future Capabilities estimated the extent of the dependency of the Naval Services on space systems—specifically on capabilities provided by National Security Space (NSS) mission areas; the committee's estimate is summarized in Figure ES.1. An assessment of the status of current and proposed major DOD space systems, given in Table ES.1, indicates the Navy's reliance on space system development and operations led by other agencies. While the capabilities of many of today's space systems (in navigation, ISR, communications, and space control) originated in numerous efforts previously supported by the Navy, today most space systems supporting naval operations are operated by other government agencies or by commercial firms.

The continual replacement and the upgrading of most of these major space systems (for which the satellites typically have a design lifetime of less than 15 years) are being planned, and the deployment of these satellites is anticipated to cost tens of billions of dollars over the next 15 years. To help the DOD leverage the resulting opportunities, as well as ensuring its future needs will be met, William Cohen (then Secretary of Defense), in 2000, commissioned a space review panel to investigate NSS management and organization.[7] Shortly after becoming Secretary of Defense, Donald Rumsfeld, who had initially chaired the DOD Space Commission, enacted many of the report's recommendations. This action, promulgated as Directive 5101.2,[8] establishes a DOD Executive Agent for Space, a responsibility currently assigned to the Under Secretary of the Air Force,

---

[4]ADM Vern Clark, USN, Chief of Naval Operations; and Gen Michael W. Hagee, USMC, Commandant of the Marine Corps. 2003. *Naval Operating Concept for Joint Operations*, Department of the Navy, Washington, D.C., September 22.

[5]ADM Vern Clark, USN, Chief of Naval Operations. 2002. "Sea Power 21," *U.S. Naval Institute Proceedings*, Vol. 128, No. 10, pp. 32-41.

[6]Gen James L. Jones, USMC, Commandant of the Marine Corps. 1999. *Marine Corps Strategy 21*, Department of the Navy, Washington, D.C., July.

[7]Commission to Assess United States National Security Space Management and Organization. 2001. *Report of the Commission to Assess United States National Security Space Management and Organization*, Washington, D.C., January 11.

[8]The full text of DOD Directive 5101.2, "DOD Executive Agent for Space," is provided in Appendix B.

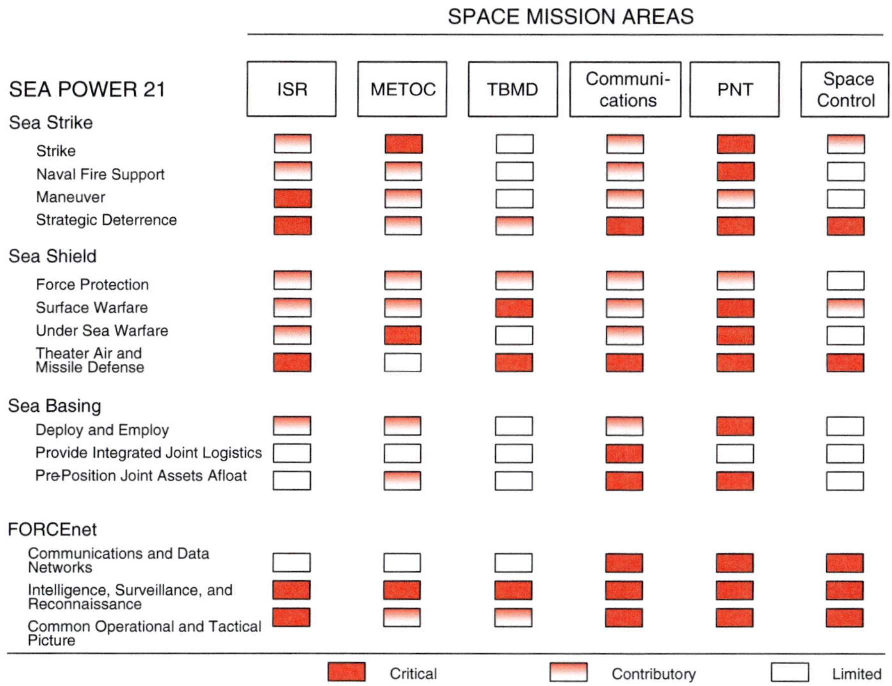

FIGURE ES.1 Dependency of Sea Power 21 as an operational framework on capabilities provided by National Security Space mission areas. (A list of acronyms is provided in Appendix G.) NOTE: A "critical" dependency reflects a space mission area that is considered absolutely necessary for the accomplishment of the particular Sea Power 21 capability. For example, without access to space-based ISR information, the Navy could not generate the common operational and tactical picture needed to support FORCEnet; thus, these two areas are denoted critical. A "contributory" dependency reflects a space mission area that will provide support for accomplishing the particular Sea Power 21 capability.

and directs the Services to assist the DOD Executive Agent for Space in developing capabilities and priorities to meet their needs.[9]

The DOD Executive Agent for Space is specifically tasked to collect input from the relevant Service and intelligence communities in order to coordinate,

[9]In addition to establishing the DOD Executive Agent for Space, the Secretary of Defense also recently disestablished the U.S. Space Command and assigned its responsibilities to the U.S. Strategic Command. The three Service-component commands are the Air Force Space Command, the Army Space and Missile Command, and the Naval Network Warfare Command. The Marine Corps Strategic Command was also recently established to provide direct Marine Corps input to the U.S. Strategic Command.

TABLE ES.1 Status of Current, Planned, and Proposed Department of Defense Space Systems Programs, Including General System Limitations and Risk

Space-Based Capability	Current Programs	Available?	Oversight Agency
Imagery (infrared, visible, radar)	NRO systems, FIA, SBR commercial imagery	Yes	DOD Executive Agent for Space, NRO, NGA
Electronic intelligence (ELINT)	NRO systems	Yes	DOD Executive Agent for Space/NRO
Navigation	GPS	Yes	Air Force
Timing	GPS	Yes	Air Force/Navy
Meteorology and oceanography	GOES, POES, NPOESS	Yes	NOAA
Ground moving target indication	SBR	No	Air Force
Airborne moving target indication	None	No	None designated
Boost-phase missile defense	SBIRS-H	No	MDA
Midcourse missile defense	SBIRS-L	No	MDA
Space-based IP networks (GIG)	TCA	No	DOD Executive Agent for Space
Satellite communications	MUOS, MILSTAR, AEHF, commercial	Yes	Air Force and Navy

NOTE: A list of acronyms is provided in Appendix G.

prioritize, program, acquire, and operate all National Security Space systems. Not included among the NSS systems are the nation's meteorology and oceanography systems, which are currently mandated to fall under oversight by the National Oceanic and Atmospheric Administration (NOAA) as executive agent.

In light of these organizational changes, the Navy should revisit its approach to supporting and leveraging national space capabilities. Only two current space systems, the Mobile User Objective System (MUOS) and the Geodetic Satellite (Geosat), are currently under Navy control. Some in the Navy have argued that a passive stance can be taken—that it is the Air Force's responsibility to supply all space-related military needs and that the Navy should act only as a "ruthless customer" of NSS capabilities. In this view, useful naval adaptations, once systems are on-orbit and in operation, could be made through the Navy's Technical Exploitation of National Capabilities Program (Navy-TENCAP), as has been done in the past. Others in the Navy indicated to the committee their concern that

General Limitations	Status/Risk
Satellite revisit time	FIA and SBR in development, SAR versus GMTI trade-offs for SBR
Encryption, geolocation accuracy	Gap-filler satellite may fill need
Vulnerability to countermeasures	Enhanced jamming protection programmed
Few	Ongoing clock development
Passive sensors only, resolution and revisit times, international partnerships	Cost, feasibility of active sensors
Revisit time, field of view, data rates	R&D issues, cost, trade-offs between SAR and GMTI, ubiquitous coverage
Stressing technology, data rates	Ubiquitous coverage, use of space sensor for weapon guidance
Stressing technology	Cost, technical risk, system under development
Stressing technology	Cost, technical risk, no current program
Wideband laser links	Development risk, cost, system under development
Data assurance, link availability	Bandwidth demands growing rapidly

the Air Force may not take sufficient account of naval needs as system design trade-offs are made to meet cost and schedule constraints or as system priorities are modified.

The committee concluded that, in the foreseeable future, a change in the executive agent roles for the Air Force and NOAA is unlikely. Thus, a more active role should be assumed by the Navy and Marine Corps, through multiple interfaces with the executive agents, to ensure that naval space support needs are met and to take advantage of opportunities offered by the large funding outlays now being made.

In 2002 the Panel to Review Naval Space recommended several steps toward such an active role.[10] The present committee reviewed progress on these recom-

[10]Panel to Review Naval Space. 2002. *Report of the Panel to Review Naval Space: Assured Space Capabilities for Critical Mission Support*, Center for Naval Analyses; Alexandria, Va., March 19.

mended directions, concluding that more can and should be done to ensure that the naval forces' needs for support from space are met. The committee's major recommendations, presented below, are parallel to those of the 2002 review panel study, and are focused on the following areas:

- Establishing a new Department of the Navy space policy,
- Determining and articulating Navy space needs,
- Increasing participation in National Security Space activities,
- Reinvigorating support of Navy space science and technology,
- Enhancing experimentation in the development of space systems,
- Strengthening the naval space cadre, and
- Taking technical and programmatic steps to leverage National Security Space mission areas.

RECOMMENDATIONS FOR MEETING NAVAL FORCE SPACE NEEDS

Establish a New Department of the Navy Space Policy

The Department of the Navy should first move to fulfill the responsibilities assigned to it by the DOD in Directive 5101.2, that is, to assist the DOD Executive Agent for Space in developing maritime space capabilities. The committee's perception is that not much has been done by the Navy Department toward compliance with this directive. For example, to date, no documents or policy statements have been provided to the DOD Executive Agent for Space detailing how the Department of the Navy will support its assigned responsibilities.[11] In part, this may be because responsibility for this response was delegated to the Under Secretary of the Navy, and that position was vacant before the directive was promulgated.[12] Taking into consideration the lack of input from the Under Secretary of the Navy, the committee was nonetheless concerned that the Department of the Navy had not yet begun to fulfill its responsibility under DOD Directive 5101.2 by updating the Department of the Navy space policy—last done in 1993.[13]

[11] One positive step toward coordinating Navy space activities was reported in mid-April 2004; the article indicates that the Navy will create a Program Executive Office for Space: Amy Butler. 2004. "Navy to Establish Space Program Executive Officer," *Defense Daily*, April 14, p. 5.

[12] A new Under Secretary of the Navy was nominated on February 6, 2004, and confirmed on October 8, 2004, while this report was undergoing classification review.

[13] John H. Dalton, Secretary of the Navy. 1993. SECNAV Instruction 5400.39B, "Department of the Navy Space Policy," Department of the Navy, Washington, D.C., August 26.

Recommendation 1. The Secretary of the Navy should task the Chief of Naval Operations and the Commandant of the Marine Corps to formulate and take steps to establish a new Department of the Navy space policy.

This space policy should provide a framework for Department of the Navy participation in the planning, programming, and acquisition activities of the Department of Defense Executive Agent for Space, and include definition of the Navy Department's relationship to National Security Space activities. A primary objective of the Department of the Navy space policy should be to focus attention on space mission areas critical to the successful implementation of Naval Power 21 as well as on other national maritime responsibilities such as homeland defense.

Determine and Articulate Navy Space Needs

Recently the Navy has begun to alter its process for generating space-related requirements to place a greater emphasis on operational inputs and greater reliance on joint Service interoperability and integration of needs.[14] However, in the information available to it, the committee could not identify a rigorous, end-to-end, well-connected process within the Navy for determining, prioritizing, and articulating naval space needs. The contributions of space support to existing and anticipated naval missions should be analyzed operationally for costs and benefits, and requirements for such support should be formulated and strongly articulated in a timely fashion in order to have an impact at the points in the process at which system-design trade-offs are made. In general, the earlier a Navy need can be identified and its usefulness clearly articulated, the easier it becomes to support that need throughout the processes of requirements generation, prioritization, and acquisition.

The lack of an up-front, crosscutting analysis of naval force space needs appears to have led to a lack of clear requirements for the space assets and capabilities that are necessary to accomplish Sea Power 21. For example, there was insufficient operational analysis linking communications connectivity needs with ongoing ISR capability developments to ensure that naval vessels will be able to access the high-volume data sets envisioned in the near future—a capabil-

[14]These changes include the assignment of the Commander, Fleet Forces Command (CFFC) with responsibility for generating and prioritizing all operational requirements and the establishment of the Joint Capabilities and Integration Development System. Currently, operational inputs for space-related requirements provided to the CFFC are collected and prioritized by the Naval Network Warfare Command, which receives input from the Naval Network and Space Operations Command and the Information Operations and Space Integration Branch of the Strategy and Plans Division of Marine Corps Headquarters.

ity particularly needed to enable sea basing of a joint task force headquarters as envisioned in Sea Power 21.[15] Without clear requirements supported by sound analysis going forward, the Navy lacks the backing to articulate and ensure adherence to its requirements as programs progress through the Navy, joint Service, and DOD acquisition process. In addition, the lack of articulated needs appears to have kept space capability gaps from being identified and forwarded to the experimentation and science and technology (S&T) communities for further examination.

Recommendation 2. The Chief of Naval Operations (CNO) should task the appropriate organizations—including the Commander, Fleet Forces Command; the Deputy CNO for Warfare Requirements and Programs; and the Deputy CNO for Resources, Requirements, and Assessments—to strengthen the Navy's requirements process for identifying space capability needs.

Specifically, the Navy should increase its support of operations research, systems analysis, and systems engineering (both internally and externally performed), since the Navy appears to lack sufficient resources in these areas. Operational analysis is central to the process of integrating needs across Sea Power 21 capability areas and National Security Space mission areas. The results of this analysis should be articulated for purposes of prioritization to the appropriate organizations—those with responsibility for requirements, acquisition, science and technology, and experimentation. In this process, these organizations should use common simulation, modeling, and analysis tools that are also compatible with the Joint Capabilities Integration and Development System (JCIDS).[16]

Increase Participation in National Security Space Activities

In the absence of sustained, active participation by the Department of the Navy in staffing, management attention to, and funding of NSS programs, it is

---

[15] As discussed in the section entitled "Space-Based Communications" in Chapter 4 of this report, Navy plans are to increase command ship (LCC) bandwidth capacity to approximately 10.5 Mb/s by FY07, in stark contrast to joint task force bandwidth usage of up to 750 Mb/s during Operation Iraqi Freedom.

[16] The JCIDS process is based on top-level strategic direction, provided by the Joint Requirements Oversight Council (JROC), to guide development of new capabilities. Capability recommendations and requirements are developed by the Services and evaluated by the JROC in consideration of how to optimize joint force capabilities and maximize interoperability. All major new programs are expected to participate in the JCIDS process.

unrealistic to expect NSS programs to consistently take account of and support technical and operational requirements that are unique to naval needs. The committee's perception, in particular, was that the Department of the Navy maintains an uncertain posture in relation to most NSS programs.

In contrast, the long-standing Navy participation in the activities of the National Reconnaissance Office (NRO) seems effective and could serve as a model for its interactions within the NSS realm. Also, Navy-TENCAP activities seem to be well supported by the Navy and, while limited by law to the exploitation of existing (in-orbit) space systems, Navy-TENCAP activities have often led to new naval capabilities.

A current example of problems arising from the Navy's lack of participation in NSS activities is the difficulty that the Navy has been having in establishing its maritime requirements as part of the Air Force Space Based Radar (SBR) program. SBR plans are to field a single system capable of collecting both ground moving target indication (GMTI) and synthetic aperture radar (SAR) imagery; the committee believes that this system has significant potential to meet a variety of naval needs, including open-ocean surveillance. The Navy does not appear to have established the organization, technical depth, analytic ability, and funding to significantly influence SBR. As a result, the Navy appears to be struggling to keep up, resorting to last-minute nonconcurrence in decisions of the Joint Requirements Oversight Council (JROC) when programs do not properly reflect naval requirements. In addition, GMTI imagery must contend with Doppler-induced background clutter that has significantly different character for land use and for marine use, signifying a potential area in which Air Force land use priorities could impact Navy needs. Thus, a lack of early Navy support for naval interests in this area may result in not having naval space needs met, or in having to fund alternative solutions at a later date and greater expense.

Recommendation 3. The Secretary of the Navy should ensure that the naval forces are adequately staffed and supported to influence National Security Space programs that have the potential to meet important naval space needs.

The Navy should engage early, with sufficient technical and management depth to influence the requirements generation, resourcing, and acquisition of new space systems being developed by the Air Force, National Oceanic and Atmospheric Administration, National Reconnaissance Office, and other (commercial and government) partners. In addition, the Navy's engagement should include tasking appropriate naval commands to provide inputs for (see Chapter 3, "Roles and Responsibilities") and participate directly in (see Table ES.2 in the final subsection of this Executive Summary) NSS activities.

Reinvigorate Support of Navy Space Science and Technology

The DOD Space Commission report,[17] as well as the related DOD directives issued subsequently,[18] emphasized the vital importance of U.S. leadership in space. One way that this leadership can be assessed is through the technological innovation and research and development (R&D) aimed at generating satellites that have the greatest long-term benefits.[19] However, this committee's perception is that the Navy, which at one time held a leadership position in national space science and technology, has allowed its support of space S&T to wither to such an extent that it is no longer adequate to support Navy needs for broad expertise in space issues or the technical expertise in space-based ISR, communications, and METOC systems essential to Sea Power 21.[20] The Navy's Mobile User Objective System program is an exception: enduring maritime needs for mobile, narrowband satellite communications have justified continual Navy support of ultrahigh frequency communications and signal propagation research. While necessary to support MUOS, this R&D effort does not generate broad technical knowledge across other NSS mission areas and may result in the naval requirements community's overlooking opportunities to increase Navy capabilities using space systems.

The committee has a particular concern for the Navy's in-house base of space technology expertise: the Naval Research Laboratory's (NRL's) Naval Center for Space Technology (NCST). Beyond the "core" support provided from NRL, there has been little recent support for NCST by the Office of Naval Research (ONR); rather, funding from non-naval organizations has allowed NCST to maintain a critical mass of personnel, facilities, and technical credibility to support the development of a number of capabilities useful to the Navy.[21] NCST staff described to the committee several proposals to develop novel maritime space capabilities (for direct user tasking, emerging wideband communications to disadvantaged naval platforms, and improved radio-frequency emitter tracking)

---

[17]Commission to Assess United States National Security Space Management and Organization. 2001. *Report of the Commission to Assess United States National Security Space Management and Organization*, Washington, D.C., January 11.

[18]DOD Directive 5101.2, "DOD Executive Agent for Space," with an enclosure listing other relevant directives, is presented in Appendix B of this report.

[19]Center for Strategic and International Studies (CSIS) Satellite Commission. 2002. *Preserving America's Strength in Satellite Technology*, CSIS, Washington, D.C., April.

[20]For example, Office of Naval Research (ONR) space-related funding in environmental effects and spacecraft technology research was approximately $14 million in FY03—less than 1 percent of total ONR FY03 funding, according to RADM Jay Cohen, USN, Chief of Naval Research, presentation to the committee, October 29, 2003.

[21]One recent example is NCST's construction of the WindSat sensor, used to measure sea-surface wind speed and direction, and the integration of the sensor onboard the recently launched Coriolis satellite.

that have since received support from the DOD through the TacSat program.[22] The committee finds that the Navy does not appear to be interacting with NCST and taking advantage of NCST's potential to develop maritime space mission concepts, to demonstrate on-orbit space-based capabilities that support maritime operations, or to transition technology to industrial or government partners for system production.

Several emerging technologies, such as space-based radar and hyperspectral imaging (for which NCST has constructed the prototype Naval EarthMap Observer (NEMO) sensor payload), can provide capabilities to support Sea Power 21. This can happen, however, only if the naval forces provide the leadership and support necessary to develop the technical understanding of how systems can be designed and integrated into new platforms and to provide the knowledge to articulate system requirements to the broader NSS community.

Recommendation 4. The Chief of Naval Research (CNR) should maintain a critical level of space mission area funding aimed at supporting current maritime needs as well as at providing broad support to base-level technologies with the potential to support National Security Space programs, such as the Transformational Communications Architecture and Space Based Radar programs.

Specifically, the CNR should continue (or preferably increase) current levels of basic research (6.1) and applied research (6.2) funds in support of space technologies and systems. In addition, the CNR should consistently allocate advanced research and development (6.3) funds to enable regular Navy space sensor development and on-orbit testing. Given recent Navy space sensor program allocations as a benchmark, the committee envisions that a level on the order of $40 million annually of 6.3 support would be sufficient to ensure regular development of new Navy space systems. Specific space mission areas recommended are as follows:

- *Communications.* Robust on-orbit capabilities supporting naval communications needs such as connecting historically disadvantaged users to future sea-based command centers via the Global Information Grid and creating the low-latency weapons control connections necessary for effective missile defense;
- *Intelligence, surveillance, and reconnaissance; and meteorology and oceanography.* Improved on-orbit capabilities, ranging from modified sensors to new capabilities including hyperspectral imaging, to support emerging needs arising from Sea Power 21; and
- *Data fusion.* Using space-derived information and systems in scaling and optimizing global information capabilities in support of Sea Power 21 operations.

---

[22]Additional information on TacSat is provided in Chapter 3, in the section entitled "Navy Space Support."

Additional space mission area recommendations are provided in Table ES.2 in the final subsection of this Executive Summary.

Enhance Experimentation in the Development of Space Systems

The NSS community's ongoing transformation of space systems will continue to offer new opportunities to improve naval capabilities. One ready means for leveraging these opportunities is through experimentation. Typically, experimentation is used not only to test new technologies that may prove useful to the operational forces, but it is also used to provide a forum in which operators test developing technologies and provide feedback on how the technologies, or underlying concepts of operations, can be modified to improve capabilities. Additionally, experimentation often identifies or highlights capability gaps in need of further R&D. A key to the experimentation process, however, is the clear identification and prioritization of operational needs; only when these needs are established can appropriate, focused experimentation efforts be constructed.[23]

Current Navy experimentation programs, however, do not appear to be derived from articulated operational needs and thus may not effectively support new capabilities. In part this is because the Navy has not conducted rigorous analyses of the missions and capabilities needed to accomplish the goals of Sea Power 21 (see Recommendation 2). The Navy has acknowledged the need to tie operational needs more effectively into experimentation efforts, which has resulted in the establishment of Sea Trial.[24] Sea Trial is described as the "process to go from strategy based concepts through experimentation to proposed [mission capability plans] and the [naval capability plan], to changes in doctrine, organization, training, material, leadership development, personnel, and facilities (DOTMLPF)."[25] While led by the Commander, Fleet Forces Command (CFFC), the Sea Trial effort is managed by the Navy Warfare Development Command (NWDC), and for issues related to space and network-centric operations Sea Trial is also guided by the Naval Network Warfare Command (NETWARCOM).[26] In addition, Sea Trial is coordinated with the Joint Forces Command experimentation process, which is exploring the best uses of space-based intelligence capabilities for all of the Services.

---

[23]For a more thorough discussion of the Navy's experimentation efforts and structure, see the (2004) National Research Council report entitled *The Role of Experimentation in Building Future Naval Forces*, The National Academies Press, Washington, D.C.

[24]ADM Robert J. Natter, USN. 2003. "Sea Power 21 Series—Part VIII; Sea Trial: Enabler for a Transformed Fleet," *U.S. Naval Institute Proceedings*, Vol. 129, No. 11, pp. 62-66.

[25]Commander, U.S. Fleet Forces Command. 2003. *Sea Trial: Concept Development and Experimentation Campaign Plan (U)*, Norfolk, Va., June 30, p. 11 (an unclassified excerpt from a classified document).

[26]The Naval Network and Space Operations Command provides operational input to NETWARCOM on space-related experimentation issues.

The Navy's current experimentation programs involving space have been largely opportunistic, taking advantage of new technologies predominantly developed outside the Navy, to make incremental improvements in fleet performance. In addition, the Navy has focused heavily on advanced antenna technologies usable from large mobile platforms. Navy-TENCAP is a commendable example of experimentation with fielded NSS programs (ISR in particular), enabling those systems to provide critical intelligence to the tactical users. Today, a benchmark for Navy tactical exploitation of space is the Navy's use of the national electronic intelligence (ELINT) systems.

Current Navy space system experimentation does not appear to have come from a strategic and tactical planning process that identified and prioritized new capabilities necessary to meet the goals of Sea Power 21 successfully. As a result, there appears to have been limited success in transitioning successful space experimentation results into programs of record. To improve this situation, the committee believes that a closer integration among CFFC, NWDC, and NETWARCOM is needed.

Recommendation 5. As part of the Sea Trial experimentation process, the Commander, Fleet Forces Command, should formalize the roles between the Naval Network Warfare Command (NETWARCOM) and the Navy Warfare Development Command (NWDC) pertaining to maritime and joint forces experimentation in space and space-related areas so as to fully exploit and complement the Joint Forces Command experimentation process and to explore the best uses of future space-based intelligence capabilities.

In particular, NETWARCOM and NWDC should carry out the following:

- Coordinate with the Deputy Chief of Naval Operations for Warfare Requirements and Programs in order to generate experimentation initiatives aimed at addressing space capabilities requirements;
- Perform analysis, modeling, and simulation in a simulation-based acquisition approach on potential new space capabilities before proceeding to testbeds and field experiments; and
- Conduct experimentation aimed at supporting new or improved sensors and subsystems that can piggyback on available NSS satellites.

Strengthen the Naval Space Cadre

The current naval space cadre represents an excellent pool of uniformed Navy and Marine Corps officers and civilians, trained and experienced with existing space systems and with the use of these systems' products in tactical applications. However, recent downturns in Navy-funded space projects and a shift in space mission area responsibilities to the DOD Executive Agent for Space (and thus to the Under Secretary of the Air Force) appear to be leading to a downturn in the ability of the space cadre to retain its viability. Thus, the committee sees an expanded space cadre as the key to ensuring that naval space needs can be articulated, addressed, and satisfied. The Navy's demonstrated leadership in the NRO, MUOS, and Navy-TENCAP has relied on having educated, trained, and motivated teams of personnel knowledgeable about space to help develop new space systems.

As discussed above, emerging NSS programs, for example, SBR, offer new opportunities for naval warfighting capabilities. However, without the Department of the Navy's sustained involvement in all phases of these programs, it is unlikely that they will deliver the level of performance that the Navy will need in the future. The Naval Services recently moved to improve the development and organization of their respective space cadres. For example, the Navy established a position for a flag officer whose primary responsibility is to direct and oversee the future development of the Navy space cadre. As part of this effort, the Navy also created the first listing of space-rated billets and space-rated personnel (collectively defined as the Navy space cadre). As the Navy space cadre develops, it is anticipated that more and more space-rated billets will be filled by members of the space cadre (in 2003, only approximately 20 percent of the space billets were occupied by members of the space cadre).[27] The recently established Marine Corps space cadre is still developing, and its influence in Marine Corps space planning and programming is being established.

The underpinning for a knowledgeable space cadre starts with advanced education, such as that provided through the Naval Postgraduate School (NPS) Space Systems programs. These programs provide graduate education in space systems engineering and operations for officers of all the military services. A Joint Space Oversight Board, chaired by the DOD Executive Agent for Space, has been established to ensure consistency between efforts of the NPS and the Air Force Institute of Technology in space education. Navy leadership must remain actively engaged in this forum to ensure that both the quality and the scope of the NPS curricula remain responsive to maritime needs. The committee also noted that Marine Corps involvement with regard to the direction and content of the space systems curricula has increased, representing recognition of the growing relationship between space expertise and the needs of expeditionary warfare.

Relative to the trends of the previous decades, a disturbing decrease was noted in the recent Navy quotas and assignments to the NPS Space Systems programs.[28] Given the opportunities, challenges, and responsibilities offered un-

---

[27]According to RADM Rand H. Fisher, USN, in discussions with the committee, October 29, 2003.

[28]According to Rudolf Panholzer, chair, Space Systems Academic Group, Naval Postgraduate School, in discussions with the committee, November 14, 2003.

TABLE ES.2 Recommendations for Department of the Navy Participation in and Support of the Six National Security Space Mission Areas

Space Mission Area	Recommendation	
Intelligence, Surveillance, and Reconnaissance	4.1	The Department of the Navy should develop and fund directed operational analysis and science and technology (S&T) programs focused on addressing the Navy's intelligence, surveillance, and reconnaissance shortfalls independent of whether or not the affected programs are managed by the Department of the Navy or the Department of Defense (DOD) Executive Agent for Space. The Department of the Navy should also work to transition the results of these efforts into planned and ongoing National Security Space programs.
	4.2	The Department of the Navy should continue its full support of National Reconnaissance Office intelligence, surveillance, and reconnaissance (ISR) activities and seek to extend its involvement in ISR program planning, development, and execution across other agencies' ISR efforts.
	4.3	The Department of the Navy should provide budget authority to augment National Security Space intelligence, surveillance, and reconnaissance programs to permit program and system additions that address needs unique to Navy strategies (such as maritime operation).
	4.4	The Department of the Navy should coordinate with other agencies to support the development of advanced sensing technologies not currently part of the program plans of the DOD Executive Agent for Space. One such program that the committee believes has significant potential to provide new naval capabilities is the Naval EarthMap Observer (NEMO) hyperspectral imaging satellite.
Meteorology and Oceanography	4.5	The Department of the Navy should remain involved in developing and operating Navy-unique satellite systems. Thus, the Department of the Navy should reassess its meteorology and oceanography (METOC) remote sensing priorities. It is the view of this committee that these assessments should focus on the following: • Ensuring strong support for the Geosat (Geodetic Satellite) Follow-on program, • Completion and launch of the Naval EarthMap Observer (NEMO) satellite, and • Completion and launch of the Geosynchronous Imaging Fourier Transform Spectrometer/Indian Ocean METOC Imager (GIFTS/IOMI) satellite.
	4.6	The Department of the Navy should pursue research and development of integrated active and passive microwave satellite sensors development programs with the National

continues

TABLE ES.2 Continued

Space Mission Area		Recommendation
		Aeronautics and Space Administration (NASA) and the National Oceanic and Atmospheric Administration (NOAA) to enable all-weather meteorology and oceanography sensing, along with measurements of trafficability, fog and visibility, and sea-ice mapping. The Navy should also continue to explore other research demonstrations, including active satellite systems and higher-resolution systems for hyperspectral imaging and sounding, atmospheric refractivity characterization and prediction, ocean color and biological constituents monitoring, and denied-area shallow-water bathymetry.
	4.7	The Chief of Naval Research should modify Office of Naval Research technology transition rules to allow transition-oriented funds to support non-Navy (and non-DOD) meteorology and oceanography programs such as those fielded by NOAA and NASA.
Theater and Ballistic Missile Defense	4.8	The Navy should continue its aggressive support of the E-2C aircraft Radar Modernization program so that a fleetwide capability can be achieved as soon as feasible.
	4.9	The Department of the Navy should begin operational analysis of the cost, benefits, and requirements of a cruise and ballistic missile defense system based on a multimode missile and an airborne moving target indication (AMTI) space-based radar (SBR) system. The Department of the Navy should invest in a focused science and technology program to resolve the issues that currently render an AMTI SBR infeasible.
Communications	4.10	The Department of the Navy should increase its depth of understanding of Navy and integrated joint future communications needs.
	4.11	The Department of the Navy should fund and manage an expanded operational analysis program focused on supporting research and development in space-based communications.
	4.12	The Department of the Navy not only should support research and development programs, but also should support experimental programs aimed at supporting space-based communications.
	4.13	The Department of the Navy should direct research and development aimed at the problem of low-latency communications from space-based sensors to platforms, particularly with respect to the cueing of fast-moving targets from beyond-line-of-sight sensors and national systems. Such an activity should be done in conjunction with improvements to the Cooperative Engagement Capability as well as other missile defense efforts.

TABLE ES.2 Continued

Space Mission Area	Recommendation
	4.14 The Department of the Navy should focus more science and technology efforts on consolidated antenna and terminal configurations necessary to enable near-100-percent-reliable shipboard communications.
	4.15 The Department of the Navy should support a naval space-based communications challenge and fund its science and technology (S&T) community to aggressively anticipate potential future space-based communications requirements.
	4.16 The Department of the Navy should continue its role as lead agency for narrowband communications. The Department of the Navy should direct the Mobile User Objective System (MUOS) program to direct special attention in FY05 to ensuring that MUOS will interface effectively as an edge system in the Transformational Communications Architecture, and to harden the system, as is feasible within cost and schedule constraints, against the evolving counterspace threat environment.
	4.17 The Department of the Navy should revise its strategy of relying largely on commercial and unprotected communications during conflict. The Navy should carefully review the nature of potential threats to unprotected communications, both ground- and space-based, and take these threats into account when specifying next-generation communications needs and requirements. The Navy should also determine its core warfighting communications capability needs and should specify robust protection for these minimum capabilities to ensure adequate communications capabilities in the event of a total loss of access to commercial systems.
	4.18 The Department of the Navy should increase its personnel assignments to support the Transformational Communications Architecture program. The Department of the Navy should allocate naval personnel so that on the order of 10 to 15 percent of the total military and support staffing of this major acquisition program is represented by the Naval Services.
Position, Navigation, and Timing	4.19 The Department of the Navy should retain close ties with the Global Positioning System (GPS) Joint Program Office during the development of upgraded GPS space and ground segments. The Department of the Navy should also ensure that specific applications of integrated GPS (precision weapons systems, for example) are coupled to spacecraft capabilities that affect the resistance of these systems to radio-frequency interference (jamming). The Department of the Navy should conduct trade-off studies to determine the most cost-effective approach and strategy in developing guidance systems that rely on a combination of GPS and inertial guidance capabilities.

continues

TABLE ES.2 Continued

Space Mission Area	Recommendation	
	4.20	The Department of the Navy should initiate a GPS synchronization study similar to that being conducted by the Air Force to ensure that M-code (military-only) user equipment development is synchronized with space- and ground-segment M-code capabilities.
	4.21	The Department of the Navy should sustain support to continue research and development in the area of precision timing standards and time transfer techniques, especially for potential use in future GPS space systems.
Space Control	4.22	The Department of the Navy should explore potential sea-based space control concepts in coordination with the activities of the DOD Executive Agent for Space.

der the new DOD Executive Agent for Space, reversing this downward enrollment trend will be important.

Recommendation 6. The Chief of Naval Operations should strengthen and expand the Navy space cadre as follows:

- Continue formalizing the leadership of the Navy space cadre under the Deputy Chief of Naval Operations for Warfare Requirements and Programs;
- Provide additional (new) billets to support National Security Space (NSS) research, development, and acquisition efforts;
- Ensure opportunities for positions of responsibility in all NSS activities and space mission areas;
- Review the function of fleet and operational staff billets and assign space codes to billets as appropriate; and
- Reexamine the Navy's support of and quotas for the Naval Postgraduate School space systems programs in light of expanded naval involvement in NSS activities.

Take Technical and Programmatic Steps to Leverage National Security Space Mission Areas

In addition to reassessing large-scale naval involvement in and support of space activities, the Department of the Navy will also need to improve its approach to leveraging the unique opportunities present in each of the NSS mission areas. Table ES.2 lists the committee's technical and programmatic recommendations resulting from the detailed discussion of Navy space support presented in Chapter 4, "Implementation: Navy Support to Space Mission Areas." The recommendations listed in the table are aimed at encouraging the Department of the Navy to participate in and support each of these mission areas, taking advantage of the unique opportunities and challenges associated with each of these mission areas and advocating technical and programmatic means to ensure that these NSS areas support naval needs.

1

Introduction

THE NAVY AND SPACE, PAST AND PRESENT

The Navy and the Marine Corps both as users and as developers have a long, highly diverse, and important history involving space.[1] Every day, naval forces around the world rely on hundreds of thousands of "space exchanges"—Global Positioning System (GPS) locations, instant messages, e-mails, and weather images—all due to an amazing history of space-related developments. The Navy's interest in space is long, officially starting with the establishment of the forerunner of the U.S. Naval Observatory in 1830. Its interests in space are diverse, divided among surveillance, communications, navigation, environmental monitoring, rocket design, and support of manned spaceflight. And the Navy's interest in space is important, because without space contributions it would not be possible to conduct modern naval operations. The importance of space to naval operations is the reason for examining the history as part of the prologue to future investments in naval space research and development. This history (further expanded in Appendix A) focuses first on the significant leverage gained from satellite system performance supporting naval needs and then on the resulting naval satellite acquisition and operational efforts.

Many "firsts" in space are a credit to earlier naval space research and development activities: first space communications used in operations—Moon Bounce;

[1]An extended review of the Navy's history in space is presented in Gary Federici, Robert Hess, and Kent Pelot, 1997, *From the Sea to the Stars: A History of U.S. Navy Space and Space-Related Activities*, Working Paper, Center for Naval Analyses, Alexandria, Va., and in the references cited therein.

first controllable launch vehicle—Viking/Vanguard; first satellite tracking system—Minitrack; oldest orbiting satellite—Vanguard; first successful electronic intelligence reconnaissance satellite—GRAB (for Galactic Radiation and Background); first space navigation satellite—Transit; first space object tracking system—the Naval Space Surveillance System; first demonstration of on-orbit atomic clocks—Timation, which led to the current GPS constellation; first operational military broadcast satellite—the naval Fleet Satellite (FLTSAT) communications system; first American man in space; first American man to orbit Earth; first Space Shuttle crew; first American woman astronaut; and many others. The list is long and diverse; the results are a key enabler for today's naval operations.

These and many other technical accomplishments were advanced in a larger historical environment that included the following:

- The large U.S. space effort, urgently begun in 1958: formation of the National Aeronautics and Space Administration (NASA) (to which much of the Navy space technology capability was transferred) required Navy operational support and used Navy and Marine Corps astronauts;
- Appreciation by the U.S. Congress of the large expense of space programs, and the push for "common user systems"—for example, weather satellites;
- Issuance of the Department of Defense (DOD) directive (with the transfer of the Advanced Research Projects Agency (ARPA) space program and the formation of the National Reconnaissance Office) giving the Secretary of the Air Force the "Executive Agent" role in 1961;[2] it was rescinded in 1970[3] but reinstated in 2003;[4]
- Dominant strategic nuclear priorities in DOD space programs, which continued to the end of the Cold War;
- Large-scale, rapidly developing commercial space communications activity, beginning with the first commercial communications satellite (COMSAT) in 1961, exploited by the Navy beginning in the mid-1960s;
- Experiments by the Naval Research Laboratory with the Massachusetts Institute of Technology Lincoln Laboratory's Lincoln Experimental Satellite series in the mid-1960s, indicating advantages of ultrahigh-frequency (UHF) communications for Navy ships, which led to the Navy's proposal for a UHF fleet satellite communications constellation (later called FLTSAT);

---

[2]Robert McNamara, Secretary of Defense. 1961. "Development of Space Systems," DOD Directive 5160.32, Department of Defense, Washington, D.C., March 6.

[3]David Packard, Secretary of Defense. 1970. "Development of Space Systems," DOD Directive 5160.32 (Revised), Department of Defense, Washington, D.C., September 8.

[4]Paul Wolfowitz, Deputy Secretary of Defense. 2003. "DOD Executive Agent for Space," DOD Directive 5101.2, Department of Defense, Washington, D.C., June 3.

- Soviet fleet expansion in the 1970s, leading to Navy Outlaw Shark experiments on over-the-horizon targeting;
- Resurgent DOD space activity in the early 1980s, leading to these events: recommendations made by the Naval Studies Board for increased Navy space activity and a space cadre;[5] establishment of the Strategic Defense Initiative; establishment of a commander-in-chief for space; creation of the Navy Space Command; issuance of the first Navy space policy statement; and recognition of the Navy's role in theater and ballistic missile defense (TBMD);
- Use of Air Force Defense Satellite Program satellites to detect some types of large aircraft;
- The end of the Cold War, together with the Navy's new strategy in the early 1990s of overland power projection from the sea, leading to a new Navy space policy and an expanded need for overland space imagery for Navy fire support (especially important during operations in Kuwait and Kosovo);
- A network-centric warfare thrust in the mid- to late-1990s, indicating a need for greater Navy space connectivity;
- An erosion of Navy (and national) space science and technology capability;
- Renewed space activity by all Services initiated by Secretary of Defense Donald Rumsfeld; a new, responsive Defense Advanced Research Projects Agency (DARPA) space program established; Secretary of the Air Force (again) given the DOD oversight role for space;[6] and a panel report issued by the Center for Naval Analyses recommending renewed Navy response to space development;[7] and
- The appearance of recurrent themes of space warfare and space control (studied beginning in the late 1950s) and space-based radar (studied intensively for ocean surveillance by the Navy in the 1970s).

Over the next 6 years (FY04 through FY09), the Navy plans to spend approximately $1.3 billion on space-related activities annually.[8] Of this total, nearly 90 percent is allocated to the Navy's communications satellite programs (the

[5]National Research Council. 1980. *Volume I: Report of the Panel on the Implications of Future Space Systems for the U.S. Navy (U)*, National Academy Press, Washington, D.C. (classified).

[6]DOD Directive 5101.2 assigns the DOD Executive Agent for Space responsibility to the Secretary of the Air Force, but goes on to further delegate the responsibility to the Under Secretary of the Air Force. The complete text of DOD Directive 5101.2 is presented in Appendix B of this report.

[7]Panel to Review Naval Space. 2002. *Report of the Panel to Review Naval Space: Assured Space Capabilities for Critical Mission Support*, Center for Naval Analyses; Alexandria, Va., March 19.

[8]Office of the Under Secretary of Defense (Comptroller). 2002. "RDT&E Programs (R1)," *Department of Defense Budget, Fiscal Year 2003*, Department of Defense, Washington, D.C., February, p. N-1.

Mobile User Objective System (MUOS) and the UHF Follow-on (UFO) system) and the acquisition of satellite communications terminals. In total, the Navy's space-related budget is allocated to the following:

- Communications satellites (UFO, MUOS), 49.8 percent;
- Satellite communications terminals, 38.7 percent;
- Naval Network and Space Operations Command, 5.2 percent;
- Global Positioning System receivers and equipment, 2.8 percent;
- Spectrum management and interference reduction, 1.5 percent;
- Navy Technical Exploitation of National Capabilities Program and the Ground Moving Target Indication Advanced Concept Technology Demonstration, 1.1 percent;
- Meteorology and oceanography (satellites and operations), 0.9 percent; and
- Missile warning, 0.05 percent.[9]

In addition to these direct space-related funds, the Navy supports a small amount of space science and technology (S&T) development work, much of it performed through the Naval Research Laboratory (NRL) and its Naval Center for Space Technology (NCST). In FY03, the Office of Naval Research (ONR) funded studies of environmental effects ($10.7 million in basic research (6.1) funds) and spacecraft technology ($3.3 million in applied research (6.2) funds) but supported no advanced research and development (6.3) projects.[10] This support amounts to less than 1 percent of the Department of the Navy's FY03 combined 6.1, 6.2, and 6.3 budget.[11] These amounts contrast sharply with typical spacecraft costs. For instance, NCST recently received a contract from the Office of Force Transformation to develop and launch a new Tactical Microsatellite (TacSat) for under $15 million (approximately $9 million for the spacecraft and $6 million for the launch); if TacSat had been supported by ONR, it would have consumed ONR's entire annual space S&T budget. Recent experimental meteorology and oceanography (METOC) satellite programs (Coriolis-WindSat, Naval EarthMap Observer (NEMO), and Geosynchronous Imaging Fourier Transform Spectrometer (GIFTS)) received (or were projected to receive) a total of $50

---

[9]Office of the Under Secretary of Defense (Comptroller). 2002. *Department of Defense Budget, Fiscal Year 2003,* Department of Defense, Washington, D.C., February.

[10]RADM Jay Cohen, USN, Chief of Naval Research, presentation to the committee, July 29, 2003.

[11]DOD budget figures show that the Department of the Navy total 6.1, 6.2, and 6.3 budgets were approximately $1.61 billion in FY03. Office of the Under Secretary of Defense (Comptroller). 2002. "RDT&E Programs (R1)," *Department of Defense Budget, Fiscal Year 2003*, Department of Defense, Washington, D.C., February, p. N-1.

million to $80 million in Navy support (primarily as 6.3 funds) for satellite development construction and prelaunch testing.[12]

Current Navy responsibilities include the operation of UFO and Geodetic Satellite (Geosat) systems and the operation (coordinated with NASA and the National Oceanic and Atmospheric Administration (NOAA)) of Coriolis—a satellite, launched through the Space Test Program, that carries the NRL-built WindSat sea-surface wind speed and direction measurement sensor. The Naval Space Surveillance (NAVSPASUR) system—in continuous operation by the Navy since 1958—was recently transferred to the Air Force. The Navy also participates in planning for NOAA's National Polar-orbiting Operational Environmental Satellite System (NPOESS), due for initial launch in 2012. NASA supports space science activities at NRL and at the Applied Physics Laboratory of Johns Hopkins University, and the Air Force, through its GPS Joint Program Office, supports NRL to provide GPS clock-monitoring and orbit-calculation functions.

CROSSCUTTING THEMES

The Department of the Navy strategy and framework for transformation and implementation of the National Security Strategy is embodied in the Naval Operating Concept for Joint Operations, of which Sea Power 21[13] is an integral capstone concept. In brief, Sea Power 21 is composed of pillars—Sea Strike, Sea Shield, and Sea Basing—and a foundation, FORCEnet, that addresses the command, control, communications, computers, intelligence, surveillance, and reconnaissance (C4ISR) architecture and operations necessary to achieve Sea Power 21.

As discussed throughout this report, Sea Power 21 has critical and enduring dependencies on capabilities provided by space mission areas under the National Security Space (NSS) umbrella:[14] intelligence, surveillance, and reconnaissance; meteorology and oceanography; theater and ballistic missile defense; communications; position, navigation, and timing; and space control (see Box 1.1 for further description of these areas). It is significant that all of these space mission areas are currently undergoing transformation as new systems or major block

---

[12]Figures extracted from recent DOD budget documentation: Office of the Under Secretary of Defense (Comptroller). 2002. *Department of Defense Budget, Fiscal Year 2003*, Department of Defense, Washington, D.C., February.

[13]ADM Vern Clark, USN, Chief of Naval Operations. 2002. "Sea Power 21," *U.S. Naval Institute Proceedings*, Vol. 128, No. 10, pp. 32-41.

[14]National Security Space (NSS) is currently headed by the DOD Executive Agent for Space and includes all U.S. military and intelligence satellite systems and much of their operational support.

improvements are developed and fielded. The Navy's continuing involvement and influence in this NSS transformation will be essential. In addition, the Navy is beginning to recognize the new environment, processes, and opportunities created by the designation of a DOD Executive Agent for Space and coincident reorganizations of other DOD offices.[15]

As part of the Navy's ongoing response to its needs and priorities, the Navy has established a multitiered approach to the development of large-scale systems (such as the space mission areas referred to above). This approach focuses on high-level input to establish and direct the Navy's priorities and responsibilities across six elements of involvement:

1. Strategic guidance,
2. Needs and requirements,
3. Acquisition,
4. Science and technology,
5. Experimentation, and
6. Personnel.

These elements help to demarcate the Navy's involvement in NSS programs into a three-tiered response: (1) guidance on the Navy's strategic goals, (2) guidance on the Navy's roles and responsibilities, and (3) specific guidance on implementation, tailored to optimize the Navy's participation within each of the relevant space mission areas. Thus, top-down direction and support backed up by rigorous operations analysis across the Department of the Navy will be needed to integrate these elements with the space mission areas and related space programs to generate the desired Sea Power 21 capabilities. Recent Navy reorganizations, particularly in the Fleet Forces Command and in the offices of the Deputy Chief of Naval Operations for Warfare Requirements and Programs, appear to be moving in the right direction to meet these needs, but updates to the Department of the Navy space policy,[16] needed to provide overarching and cohesive guidance regarding space support, remain uncompleted.

---

[15]See Appendix B for the full text of DOD Directive 5101.2 establishing the DOD Executive Agent for Space. Additional reorganizations include disestablishment of the U.S. Space Command and the transfer of its responsibilities to the U.S. Strategic Command; integration of the Naval Space Command and Naval Network Operations Command into the new Naval Network and Space Operations Command and placement of this command under the newly created Naval Network Warfare Command; and creation of a Marine Corps component command (Marine Corps Strategic Command) in the U.S. Strategic Command.

[16]The current Department of the Navy space policy was established in 1993. See John H. Dalton, Secretary of the Navy. 1993. "Department of the Navy Space Policy," SECNAV Instruction 5400.39B, Department of the Navy, Washington, D.C., August 26.

BOX 1.1
National Security Space Mission Areas

Intelligence, Surveillance, and Reconnaissance

Most direct information-gathering systems of relevance to National Security Space (NSS) activities are included under the umbrella of intelligence, surveillance, and reconnaissance (ISR) functions. These include not only optical and radar imagery systems, but also electronic and radio-frequency monitoring and detection systems (broadly described as electronic intelligence, or ELINT). Historically, the Navy has taken a strong position in the development and use of national ELINT systems.

Meteorology and Oceanography

Also described as environmental monitoring, the space mission area of meteorology and oceanography (METOC) encompasses all measurements aimed at providing information about the physical environment (air, land, sea surface, and undersea) that may be needed by the Services. These functions include atmospheric and undersea weather predictions, and they also assist Department of Defense (DOD) mapping, charting, and geodesy activities. The Navy has always led the DOD activities with respect to all over-water environmental monitoring. As a result of a 1994 Presidential Directive, all DOD and civilian environmental satellites are to transition their fielding and operation to the National Oceanic and Atmospheric Administration (NOAA) by 2010.[1] NOAA is currently the lead acquisition authority for the national Geostationary Operational Environmental Satellite (GOES) and Polar Operational Environmental Satellite (POES), while the Air Force operates the Defense Meteorological Satellite Program. The Navy has built and is still operating the Geodetic Satellite (Geosat) and its successors; data from these satellites are made publicly available through NOAA.

Theater and Ballistic Missile Defense

Early warning for theater and ballistic missile defense encompasses those systems and capabilities necessary to detect and track cruise and ballistic missile threats. Owing to the high speed of these threats, integrated low-latency, early-detection systems are critical, to provide an ability to mount a counterstrike against an incoming missile. Currently all missile defense activities (theater and strategic) are managed by the Missile Defense Agency, with the Navy assigned as the lead Service with respect to sea-based theater missile defense.

[1] National Science and Technology Council. 1994. "Convergence of the U.S.-Polar-Orbiting Operation Environmental Satellite Systems," Presidential Decision Directive, Washington, D.C., May.

Communications

Communications provide the critical command-and-control links that enable effective coordination of all the components of the DOD. Current military satellite communications are provided by a wide range of systems, most of which are fielded under the Air Force's acquisition lead. The exception is the Navy's current acquisition and operation of the ultrahigh-frequency (UHF) family of satellites (UHF Follow-on, or UFO, and the Mobile User Objective System (MUOS)).

Position, Navigation, and Timing

Space-based navigation systems enable the determination of precise, three-dimensional global positioning data as well as providing a consistent global standard timing signal. Currently these capabilities are provided by the Navigation Satellite Timing and Ranging/Global Positioning System (NAVSTAR/GPS) satellite constellation. GPS is managed by the Interagency GPS Executive Board, co-chaired by officials from the DOD and the Department of Transportation. The Air Force is currently the lead acquisition authority for NAVSTAR/GPS; the Navy is the lead Service for DOD time standards.

Space Control

The DOD defines space control as "combat and combat support operations to ensure freedom of action in space for the United States and its allies and, when directed, deny an adversary freedom of action in space."[2] Accordingly, space control encompasses three mission needs: (1) monitoring of the space environment and disposition of all space assets (national and foreign), (2) assurance that U.S. forces will have access to space-dependent capabilities even in the event of hostile action against U.S. space assets, and (3) the ability to deny adversaries access to their space assets. Current space control activities are being undertaken throughout the DOD. The Air Force Space and Missile Systems Center has established a Space Superiority System Program Office, and the Defense Advanced Research Projects Agency is also involved in the development of new space control technologies and systems. The Navy built the Naval Space Surveillance (NAVSPASUR) system that was transferred to the Air Force in 2003. Following this transfer and additional Air Force study, the Air Force decided that NAVSPASUR fills a critical space control need and will continue its operations for the foreseeable future.

---

[2]William Cohen, Secretary of Defense. 1999. "Department of Defense Space Policy," DOD Directive 3100.10, Department of Defense, Washington, D.C., July 9, p. 23.

ORGANIZATION OF THIS REPORT

The Committee on the Navy's Needs in Space for Providing Future Capabilities focused its findings and recommendations not only to provide specific recommendations for improvement with respect to space mission areas, but to also integrate these recommendations in order to offer guidance for general coordination and interaction on each of the six elements of involvement listed above. To this end, the report presents three levels of discussion: Chapter 2 addresses the space strategic framework, in which is set the Navy's vision and concepts of operations regarding support of future space systems; Chapter 3 discusses the Navy's roles and responsibilities necessary to achieve and support a Navy space strategy; and Chapter 4 provides implementation guidance tailored to the Navy's participation within each of the NSS mission areas. Chapter 5 then provides a view of the potential role for space capabilities supporting naval forces in the future and a vision of the Navy's engagement across the activities of NSS.

2

Strategic Framework: Future Operational Concepts and Space Needs

STRATEGIC ENVIRONMENT

The role of space in naval planning, training, and operations has been the subject of many previous studies for the Department of the Navy.[1] These studies focused on issues such as the development and maintenance of a naval space cadre, the adequacy of space system science and technology (S&T) funding, and the development of a coherent naval space policy. The approach of the Department of Defense (DOD) to space was the subject of a major study by a commission led by Donald Rumsfeld who, named Secretary of Defense shortly after publication of the commission's report, has now implemented many of its recommendations.[2] With the resulting transitional situation regarding DOD space activities, it is particularly noticeable that the Department of the Navy has not acted to update its policy toward its support of space technologies and systems and does not appear to be engaging in sufficient analysis and prioritization of needs to enable its effective participation in space-related efforts across the DOD.

Recent combat experience in Iraq and Afghanistan has highlighted specific trends in modern warfare that must be addressed and accommodated in ongoing

---

[1]These studies include the following: (1) National Reconnaissance Office and Department of the Navy. 1998. *Report of the Panel to Review Naval Participation in the National Reconnaissance Office*, Washington, D.C., February 17; (2) Panel to Review Naval Space. 2002. *Report of the Panel to Review Naval Space: Assured Space Capabilities for Critical Mission Support*, Center for Naval Analyses, Alexandria, Va., March 19.

[2]Commission to Assess United States National Security Space Management and Organization. 2001. *Report of the Commission to Assess United States National Security Space Management and Organization*, Washington, D.C., January 11.

naval transformational efforts. Some of these trends are the result of external factors—global terrorism and the replacement of the traditional geopolitical adversary with a number of separate threat areas, the increased importance of the littorals versus the deep ocean in naval operations, and the increased importance of sea-based interdiction versus traditional naval combat—while other trends are related to Navy changes in response to the evolving expectations and role of military forces in conflict and post-conflict situations.

In response to these trends, the U.S. approach to military actions has moved toward joint operations, increasing the use of Special Operations Forces, increasing the precision and decreasing the response time of strikes, and using effects-based and agile planning of military operations. In extreme cases, operations seem to emulate the Apollo "Mission Control" model, in which a relatively small number of ground troops are supported, coordinated, and sometimes commanded by a large staff of command, control, communications, computers, intelligence, surveillance, and reconnaissance (C4ISR) experts. Overlaying these trends is the increasing rate of development of technology, which in many cases outpaces even the fastest defense planning and acquisition efforts. This problem is especially severe in the case of space systems, where planning and acquisition cycles are traditionally very long (10 to 15 years is not uncommon). The recent trends, both operational and programmatic, coupled with the significant cost associated with space systems, led to the designation of a DOD Executive Agent for Space to oversee all National Security Space (NSS) program development.

The recently updated National Security Strategy[3] provides part of the strategic framework within which this committee examined the Department of the Navy's evolving transformational efforts and naval reliance on capabilities provided by NSS systems. The other major part of the framework employed by the committee is the combined Navy and Marine Corps vision as expressed in the Naval Operating Concept for Joint Operations.[4] The NOC, as it is known, was drafted to expand on the preceding naval vision, Naval Power 21,[5] and thus to strengthen Naval Power 21's integration of the Naval Services' capstone concepts Sea Power 21[6] and Marine Corps Strategy 21.[7] All of these documents build

---

[3]President George W. Bush. 2002. *The National Security Strategy of the United States of America*, The White House, Washington, D.C., September 17.

[4]ADM Vern Clark, USN, Chief of Naval Operations; and Gen Michael W. Hagee, USMC, Commandant of the Marine Corps. 2003. *Naval Operating Concept for Joint Operations*, Department of the Navy, Washington, D.C., September 22.

[5]Gordon England, Secretary of the Navy; ADM Vern Clark, USN, Chief of Naval Operations; and Gen James L. Jones, USMC, Commandant of the Marine Corps. 2002. *Naval Power 21. . . A Naval Vision*, Department of the Navy, Washington, D.C., October.

[6]ADM Vern Clark, USN, Chief of Naval Operations. 2002. "Sea Power 21," *U.S. Naval Institute Proceedings*, Vol. 128, No. 10, pp. 32-41.

[7]Gen James L. Jones, USMC, Commandant of the Marine Corps. 1999. *Marine Corps Strategy 21*, Department of the Navy, Washington, D.C., July.

on an earlier strategic framework, set forth in the U.S. Joint Chiefs of Staff Joint Vision 2020, which states:

> Three aspects of the world of 2020 have significant implications for the US Armed Forces. First, the United States will continue to have global interests and be engaged with a variety of regional actors. Transportation, communications, and information technology will continue to evolve and foster expanded economic ties and awareness of international events. Our security and economic interests, as well as our political values, will provide the impetus for engagement with international partners. The joint force of 2020 must be prepared to "win" across the full range of military operations in any part of the world, to operate with multinational forces, and to coordinate military operations, as necessary, with government agencies and international organizations.
>
> Second, potential adversaries will have access to the global commercial industrial base and much of the same technology as the US military. We will not necessarily sustain a wide technological advantage over our adversaries in all areas. Increased availability of commercial satellites, digital communications, and the public internet all give adversaries new capabilities at a relatively low cost. We should not expect opponents in 2020 to fight with strictly "industrial age" tools. Our advantage must, therefore, come from leaders, people, doctrine, organizations, and training that enables us to take advantage of technology to achieve superior warfighting effectiveness.
>
> Third, we should expect potential adversaries to adapt as our capabilities evolve. We have superior conventional warfighting capabilities and effective nuclear deterrence today, but this favorable military balance is not static. In the face of such strong capabilities, the appeal of asymmetric approaches and the focus on the development of niche capabilities will increase. By developing and using approaches that avoid US strengths and exploit potential vulnerabilities using significantly different methods of operation, adversaries will attempt to create conditions that effectively delay, deter, or counter the application of US military capabilities.[8]

Until the early 1990s, Cold War scenarios formed the basis for U.S. military planning and doctrine. Military planning envisaged that U.S. forces might become engaged in a major war with a nuclear-capable peer competitor. Such a competitor was also assumed to possess competent space-based intelligence, surveillance, and reconnaissance (ISR) resources and was assumed to be capable of matching U.S. forces in firepower and technological innovation. For naval forces, the Cold War threat focused on defense against submarine attacks on U.S. or North Atlantic Treaty Organization sea-based lines of supply, massed cruise missile attacks on U.S. surface forces, and ballistic missile (nuclear) attacks on the U.S. homeland and on deployed logistic bases.

---

[8]U.S. Joint Chiefs of Staff. 2000. *Joint Vision 2020*, Department of Defense, Washington, D.C., p. 4-5.

With the end of the Cold War, no single peer or near-peer competitor (with the possible future exception of China) is perceived to exist. However, many examples of dangerous military technology have proliferated worldwide and to some degree will be available to potential adversaries who might be encountered in regional or local conflicts. The most probable types of military interventions in which the United States might be involved during the next 25 to 30 years include these:

- Combat missions:
  - —Intervention for the protection of allies,
  - —Preemptive or punitive strikes in response to terrorism, and
  - —Preemptive or politically coercive actions;
- Noncombat missions:
  - —Blockade and sanction enforcement operations,
  - —Noncombat evacuation operations,
  - —Antimigration or migration support operations,
  - —Resource protection operations, and
  - —Refugee support and disaster recovery operations.

All of these potential missions will be coupled with political pressures and potential antiaccess and asymmetric threats, generating the specific missions that the future naval forces will need to contend with. This diversity of potential missions (and threats) is discussed briefly in the subsections below, with reference to how capabilities derived from the space mission areas can be used to enhance naval operations.

Information

Most of the missions described below rely on space assets to provide information—and more importantly, access to information—to enable those carrying out the missions to act more responsibly and quickly in the face of continually changing threats and conditions. These information needs are broadly characterized as a need for information dominance: that is, U.S. space-based sensors, augmented by both archival information and information derived from sensors on surface ships, submarines, manned aircraft, unmanned aerial vehicles (UAVs), unattended ground sensors, and ground observers, can allow commanders access to a significant knowledge advantage over the adversary. The dominance will be particularly pronounced if current sensor information is analyzed and assessed with low latency and if a secure, wide-bandwidth network exists that allows rapid dissemination of the derived data to relevant combat commands. Although the capabilities of U.S. space-based and other sensors are indeed impressive, they are inherently incapable of providing total information.

Total information dominance implies that U.S. forces have complete information concerning the capabilities, intent, and plan of action of hostile forces, while adversaries have no reciprocal information. Such a situation is never likely to exist. U.S. forces operate in an open society. Even if precise official statements of U.S. forces capabilities and intent are not broadcast or published, U.S. and foreign media certainly provide reasonably accurate and detailed speculations about such things. In addition, the United States does not have a monopoly on the availability of space-based sensors or long-haul communications networks. When a particular adversary of the United States does not have indigenous space-based sensors, third parties are often available to sell the desired information, or in some cases the hostile nation can seek to steal or otherwise access U.S. information sources and systems. Finally, most military operations today are conducted with other nations as coalition partners. Coordination needs among these partners create their own set of challenges with regard to the control and dissemination of information.[9]

It must be assumed that information dominance will always be relative and never complete, since adversaries will always be improving their capabilities independent of the improvements by U.S. forces. Thus, the United States will need to continually develop systems with finer granularity, enhanced resolution, and minimum latency between detection and dissemination of analyzed and assessed data. U.S. forces will also need to be capable of increasing their relative information dominance over adversaries by the denial, degradation, deception, and exploitation of the adversary's access to their information networks and their sensors. Such capabilities are variously called, or subsumed under the titles of, offensive information operations, space control, or electronic countermeasures. It is clear that in the future, U.S. forces will devote significant effort to the development of such capabilities, however designated, that effectively render the adversary both blind and deaf.

Asymmetrical Threats and Unconventional Opponents

In future years, naval forces will likely have to contend with many types of asymmetrical threats and/or unconventional opponents. The list might include but is not limited to the following:

- Terrorists;
- Drug cartels and organized criminal groups;

---

[9]Current DOD practice to solve issues related to coalition interoperability has typically involved coalition use of the same commercial satellite communications systems or loaning coalition commanders the use of U.S. ground stations. Issues related to communications standards and the like are being addressed as part of the broader Navy FORCEnet initiative and were not considered during this study.

- Hostile populations in urban areas;
- Suicide bombers;
- Hackers, communications jammers, and perpetrators of other sophisticated forms of electronic attack;
- Small, unmanned (remotely controlled) explosive-carrying platforms (UAV, surface, subsurface); and
- Chemical and biological weapons.

Many variants of asymmetrical threats and unconventional opponents may develop, seemingly limited only by the ingenuity of the attacker. As an example, terrorists, drug cartels, and organized criminal groups might undertake large-scale attacks on the families and dependents of deployed Service personnel. The attack on Khobar Towers in Saudi Arabia in 1996 has been considered a limited example of such a campaign. Kidnapping of Americans overseas has been another technique employed by terrorists (although, to date, only on a relatively minor scale). Attempts might be made to poison water supplies, degrade fuel reserves, or release biological agents in the vicinity of naval vessels in U.S. or foreign ports or in the vicinity of deployed U.S. ground combat forces.

For instance, the Navy has witnessed the sea-based suicide attack on the USS *Cole*, and, in Somalia, U.S. forces encountered a hostile population that was used as a shield by indigenous armed personnel as they conducted attacks on U.S. forces. The tactics in Somalia were moderately successful in the sense that they limited the use of U.S. firepower against the desired targets. More sophisticated and effective use of hostile civilian personnel to limit U.S. military actions can certainly be imagined. For situations in which defensive actions by U.S. naval forces limit the probability of success by suicide attackers, hostile forces may employ a large variety of small, unmanned platforms (air, surface, or underwater) to deliver explosive, chemical, or biological warheads.

Additional asymmetric threats are based on the worldwide proliferation of modern computer and communications capabilities. These capabilities can lead to use by hackers, communications jammers, and others employing sophisticated forms of electronic attack against naval forces, particularly against deployed ground combat forces. During the British conflict with the Irish Republican Army in Northern Ireland, such electronic techniques were highly developed and highly effective.

In addition to the asymmetric threats to U.S. naval forces discussed above, there is the potential of terrorist threats to the United States involving the introduction of weapons of mass destruction from the sea. The Navy's homeland defense role in supporting the Coast Guard near shore and conducting defensive operations offshore has not yet been fully defined. Nevertheless, it is clear that space assets will be of growing importance in this new mission, for purposes including the persistent surveillance and identification of surface vessels and the provision of reliable communications.

Space assets can play a role in countering or limiting the effectiveness of many asymmetric threats, but they alone will never eliminate such threats. In particular, NSS sensors can be used to intercept electronic communications among threat groups and can thereby locate the adversaries' targets and operational objectives, but as effective as they are, such sensors are far from perfect (e.g., adversaries can use encrypted data formats unbreakable by current technologies). Similarly, NSS sensors that track surface vessels need to be backed up by units capable of directly interdicting hostile ships before they reach threatening positions. There is little ongoing programmatic activity oriented to the development of future space sensors that will be specifically focused on asymmetric or unconventional naval threats.

Antiaccess Technologies

Traditionally, naval forces have faced attempts to limit their ability to exert military influence on events ashore. In World War II, such attempts were never successful. However, U.S. forces generally paid a very high price to achieve their objectives.

The antiaccess technologies (or threats of concern) to naval forces today include the following:

- Precision-guided, low radar cross-section (RCS) cruise missiles;
- Space-based ISR assets;
- Information operations;
- Chemical and biological warheads;
- Sweep-resistant sea mines;
- Wireless-detonated land mines;
- Ballistic missiles; and
- Nuclear weapons.

All of these technologies are currently available to U.S. military forces and are beginning to become available on the world arms market. As these systems become more widely available, they all represent threats to the naval forces.

Political Pressures

The future use of military force by the United States to accomplish national policy will be subject to a large number of constraints that might be characterized as political pressures. These constraints will arise from concerns of people both within the United States and throughout the rest of the world. The experience of recent conflicts suggests that opposition to military action will grow rapidly whenever a general perception arises that any of the following conditions exists:

- The stated rationale for U.S. military intervention is not justified,
- U.S. casualties are (or may be) excessive,
- Civilian casualties and collateral damage are excessive, or
- Military action is open-ended, with no plan for its conclusion.

These concerns of people worldwide, including the U.S. public, are well recognized by U.S. civil and military leadership. The first of these concerns can only be addressed by extensive public debate within the United States and by the avoidance of military interventions that have questionable rationales. The remaining three concerns can be addressed by the performance of U.S. military forces and through the precision of the U.S. targeting processes and weapons delivery systems. The quicker a military action is terminated, the fewer the U.S. and civilian casualties that will be sustained, and the more politically acceptable a U.S. military action will be to the public, both in the United States and abroad.

As a result of such considerations, the U.S. military has invested heavily in the development of precision weapons and space-based target-location systems. In general, increased weapon precision allows for the use of fewer and smaller warheads than were previously needed to ensure target destruction. The use of small warheads (where feasible) then reduces the likelihood of collateral damage.

Although precision navigation and guidance systems may limit collateral damage, many other factors come into consideration in planning military operations. Space-based sensors may be used to identify and locate targets for military weapons. However, they inherently fail to indicate the significance of the destruction of a target, and they generally do not provide any information on the time required to repair or replace a destroyed target. Target identification and location by space-based sensors must be supplemented by an understanding of the role that an individual target plays in the adversary's military and civil infrastructure as well as an understanding of the impact of the target's destruction on the termination of hostilities.

Space-based ISR can provide crucial information during the pre-hostilities stage in any theater. In such situations, space-based ISR can be used not only to provide the knowledge necessary to help develop the U.S. rationale regarding a potential conflict, but it can also provide information on an adversary's capabilities and plans.

Noncombat Missions

In noncombat operations—for example, blockade and sanction enforcement, noncombat evacuation, antimigration or migration support, resource (shipping) protection, refugee support, and disaster recovery operations—naval forces may be directed to protect the transfer of assets, people, and materials or to secure the land areas and shipping routes necessary to ensure operational success. In many of these cases, materials or people will need protection during transit, demanding a classical convoy capability to provide antisubmarine warfare and/or antiship cruise missile defense when operating near hostile areas. Space support requirements for such at-sea protection operations include all of the necessary components for the execution of combat operations. In particular though, these operations rely more heavily on broad sea control and tracking of potential hostile threats. Such information will be derived from a combination of national and organic ISR data and compiled with reach-back intelligence assets to forecast threats appropriately.

In addition, many noncombatant operations such as sanction enforcement and evacuation operations will require great precision in the identification and location of all hostile parties so that U.S. involvement can be avoided (or lessened) and collateral or unintended damage minimized. The ISR data required for such an operation generally is provided by national space assets and supplemented, as needed, by sea-based surveillance aircraft. In many circumstances, knowledge of detailed local environmental conditions (offshore winds, currents and surf conditions, visibility, land trafficability, and so on) may be the key to success. Naval forces will depend on the products of national meteorology and oceanography (METOC) satellites and ISR satellites for such data.

Combat Missions

In general, combat missions (such as intervention for the protection of allies, preemptive or punitive strikes in response to terrorism, or preemptive or politically coercive actions) rely on a similar set of space-based capabilities—chief among them, space-based, wide-bandwidth, beyond-line-of-sight (BLOS) communications. Without these capabilities, distributed units would no longer have access to information and to decision makers. These communications systems are needed not only to enable U.S. and allied forces to collaborate, but they also allow for the transmission of remote sensing information (electro-optic/infrared (EO/IR), electronic intelligence (ELINT), radar, and so on) collected by space and terrestrial assets. In addition, nearly all military units (and even many satellites) base their positioning information on Global Positioning System (GPS)-derived coordinates. Thus, access to GPS signals is a growing priority for all operations.

In general, strike missions need a synthesis of terrestrial and space assets to enable these capabilities: ground, air, and sea target detection, location, and identification; multisensor synthesis to support the suppression of enemy air defenses; weapons guidance; distributed sensor integration in support of countermine warfare; integration of distributed antisubmarine warfare sensors; synthesis and cueing of airborne moving target indication (AMTI) radar for cruise missile defense; low-latency decision making to support timely ballistic missile defense; environmental support; and bomb damage assessment.

Missions involved with engagement of terrorist groups are particularly challenging. Although naval forces have significant organic capabilities for ELINT collection against terrorists, they are aided substantially by space-based sensors that provide persistent reconnaissance to survey, identify, geolocate, and track terrorist training sites, weapons storage areas, and even individual terrorists. The interception of communications is particularly critical to enable the identification and tracking of specific terrorists and to enable the discovery of the plans of action, targets, and techniques of terrorist groups.

SEA POWER 21 AND ITS RELIANCE ON SPACE

To counter the many and varied threats mentioned here, the Navy and Marine Corps envisage that in the future they must continue to operate effectively as a forward-postured, immediately deployable force in joint and multinational environments. As stated in the NOC: "The Service visions, *Sea Power 21* and *Marine Corps Strategy 21*, recognize the challenges posed by a changing security environment and point the way to the future. The Navy and Marine Corps will leverage and integrate their respective strengths to produce a more effective and efficient Naval Force with improved warfighting capabilities for the Joint Force. The Naval Services will organize, deploy, employ, and sustain forces to conduct operations guided by the interrelated and complementary concepts of *Sea Strike*, *Sea Shield*, and *Sea Basing*,"[10] and its foundation FORCEnet.

Through their recently promulgated NOC, the Navy and Marine Corps have declared their commitment to the development of naval visions and concepts. In particular, the NOC clarifies how the Navy's capstone concept of Sea Power 21 (containing Sea Strike, envisioned to project offensive power; Sea Shield, envisioned to project defensive assurance; Sea Basing, envisioned to project operational independence; and FORCEnet, envisioned as the enabling concept for integrating warriors, sensors, weapons, networks, and platforms from seabed to space) contains and integrates components of the Marine Corps capstone concepts, Expeditionary Maneuver Warfare (EMW), Operational Maneuver From the Sea (OMFTS), and Ship-to-Objective Maneuver (STOM). In particular, OMFTS and STOM are described as integral components of Sea Strike. These concepts are further defined in Box 2.1.

In the following subsections, the four Sea Power 21 capabilities are introduced in detail, along with a description of the space mission areas and specific space-derived information and capabilities that are deemed critical to enable Sea Power 21.

---

[10]ADM Vern Clark, USN, Chief of Naval Operations; and Gen Michael W. Hagee, USMC, Commandant of the Marine Corps. 2003. *Naval Operating Concept for Joint Operations*, Department of the Navy, Washington, D.C., September 22, p. 3.

BOX 2.1
Definitions of Naval Operational Concepts for Joint Operations

Described below are key naval operating concepts as presented in the Naval Operating Concept for Joint Operations.[1]

- *Sea Strike* is a broadened concept for projecting precise and persistent naval offensive power. It describes how 21st-century naval forces will exert direct, decisive, and sustained influence in joint campaigns through the application of persistent intelligence, surveillance, and reconnaissance (ISR), time-sensitive strike, *Ship-to Objective Maneuver (STOM)*, and information operations (IO) to deliver accurate and devastating combat power.
- *Sea Shield* is a concept that describes the manner in which naval forces will protect our national interests with layered global defensive power. It is based on our sustained forward presence, and on our abilities to dominate the seas and to provide distributed and networked intelligence to enhance homeland defense, assure access to contested littorals, and project defensive power deep inland.
- *Sea Basing* serves as the foundation from which offensive and defensive power are projected, making *Sea Strike* and *Sea Shield* realities. It describes the projection, sustainment, and operational maneuver of sovereign, distributed, networked forces operating globally from the sea. *Sea Basing* will provide Joint Force Commanders with global command and control (C2) capability and extend integrated support to other Services.
- *Expeditionary Maneuver Warfare (EMW)* will serve as the Marine Corps capstone concept for the 21st century. It is the union of Marine Corps core competencies, maneuver warfare philosophy, and expeditionary heritage.
- *Operational Maneuver from the Sea (OMFTS)* is a concept for the projection of maritime power ashore. It focuses on the operational objective using the sea as a maneuver space and pitting strength against weakness. It generates overwhelming tempo and momentum; it emphasizes intelligence, deceptions, and flexibility; and it integrates all organic, joint, and multinational assets.
- *Ship-to-Objective Maneuver (STOM)* applies the principles and tactics of maneuver warfare to the littoral battle space. It allows for conducting combined arms penetration and exploitation operations from over the horizon directly to objectives ashore without stopping to seize, defend, and build-up beachheads or landing zones.
- *FORCEnet* is the enabler of these capabilities, and the operational construct and architectural framework for Naval warfare in the information age. It will allow systems, functions, and missions to be aligned to transform situational awareness, accelerate decision making, and allow Naval Forces to greatly distribute their combat power.

---

[1]ADM Vern Clark, USN, Chief of Naval Operations, and Gen Michael W. Hagee, USMC, Commandant of the Marine Corps. 2003. *Naval Operating Concept for Joint Operations*, Washington, D.C., September 22, p. 4.

Sea Strike

Sea Strike is the overarching concept describing how "21st-century Naval Forces will exert direct, decisive, and sustained influence in joint campaigns."[11] These forces will engage adversaries through the application of four critical capability areas: strike, naval fire support, ship-to-objective maneuver, and strategic deterrence. "At its heart, Sea Strike is a broad concept for naval power projection that leverages C5ISR (command, control, communications, computers, combat systems, intelligence, surveillance, and reconnaissance), precision, stealth, information, and joint strike together. It amplifies effect-based striking power through enhanced operational tempo and distant reach."[12]

Based on the Sea Strike vision, the Navy has identified, in the four critical capability areas, 14 specific capabilities that contribute critical naval needs for enabling future strike operations (see Table 2.1). Note that Sea Strike envisions not only the delivery of ordnance on hostile targets ashore, but also the provision of fire support for allied, coalition, or U.S. combat elements, and counterstrike capability to enable strategic deterrence.

A quick inspection of the Sea Strike capabilities shows that many rely on space-based assets. In particular, most modern strike missions require accurate location information for targets as well as for sensors. These targets must then be identified, classified, and tracked using a combination of assets: NSS, airborne, maritime, and human intelligence. NSS systems and airborne systems are generally used cooperatively to support time-sensitive requirements of strike. The support of time-critical missions also requires low-latency, high-bandwidth, BLOS communications to provide the coordination pathways enabling sensor-analysis-decision maker-shooter-weapon links. Space communications have also had an increasingly important role in recent conflicts in support of forced entry and other rapid, mobile ground force operations, particularly in support of Special Operations Forces and Marine Corps operations both in littorals and deep inland. Table 2.2 summarizes the capabilities derived from the six NSS space mission areas (ISR; METOC); theater and ballistic missile defense (TBMD); communications; position, navigation, and timing (PNT); and space control) that are used to augment and enable the four Sea Strike capability areas.

Sea Strike first of all relies on ISR information from NSS sensors that can be merged with ground-, sea-, and air-based sensor data to develop fused products. These products assist in the positioning and repositioning of platforms and forces,

---

[11]ADM Vern Clark, USN, Chief of Naval Operations; and Gen Michael W. Hagee, USMC, Commandant of the Marine Corps. 2003. *Naval Operating Concept for Joint Operations*, Washington, D.C., September 22, p. 4.

[12]VADM Cutler Dawson, USN; and VADM John Nathman, USN. 2002. "Sea Power 21 Part III; Sea Strike: Projecting Persistent, Responsive, and Precise Power," *U.S. Naval Institute Proceedings*, Vol. 128, No. 12, p. 54.

TABLE 2.1 Sea Strike Capabilities and Capability Areas

Strike	Naval Fire Support	Maneuver	Strategic Deterrence
Conduct strike operations. Engage fixed land targets. Engage moving targets. Conduct special operations. Provide precision targeting. Conduct direct action. Conduct offensive information operations. Jam potential threats. Conduct network attack. Provide aircraft survivability.	Provide preparation fires. Provide close supporting fires. Provide precision supporting fires. Provide volume fires.	Project forces, and reposition forces. Assault centers of gravity and critical vulnerabilities. Conduct concurrent/ follow-on missions.	Conduct nuclear strike. Provide assured survivability.

SOURCE: RADM K.J. Cosgriff, USN, Director, Warfare Integration and Assessment, "Future Force Development," slide 9, presentation to the Naval Studies Board, Washington, D.C., November 12, 2003.

assist in the identification and assault of critical vulnerabilities and centers of gravity of enemy forces, and assess damage from strikes to permit the effective use of strike resources for conducting concurrent and follow-on strikes.

Ordnance delivery may be achieved through the use of long-range, standoff missiles (e.g., Tomahawk), manned or unmanned aircraft, organic ground force firepower, or sea-based gunfire. Targets ashore are identified and geolocated through a combination of the following:

- Space-based ISR assets (ELINT, EO/IR, and synthetic aperture radar (SAR) imagery);
- Archival imagery and intelligence databases;
- Unmanned aerial vehicles or unattended ground sensors;
- Reconnaissance aircraft;
- Ground moving target indication (GMTI) radar;
- Ground observers; and
- Airborne observers and controllers.

The capability of the Navy to support Marine Corps forced entry through the littorals is, and will continue to be, a major component of Sea Strike. Such incursions certainly are and will continue to be dependent on the availability of space-based and other systems and sensors, as presented in Table 2.2.

TABLE 2.2 Space Capabilities, Derived from Space Mission Areas, Needed to Enable Sea Strike Capability Areas

	Sea Strike Areas
Space Mission Areas	Strike
Intelligence, Surveillance, and Reconnaissance (ISR) Responsive persistent ISR (imagery and signals) of static and moving targets.	Target identification, location, and tracking. Electronic intelligence (ELINT)-enabled target identification.
Meteorology and Oceanography Continuous tactical weather prediction. Acoustic/thermal modeling of the littorals.	Steaming direction for carrier aircraft launch. Cue appropriate ISR assets depending upon weather.
Communications All information must be moved, often at beyond-line-of-sight (BLOS) ranges. Sensor-analysis-decisions-shooter-weapon links.	Air tasking orders. Precision weapon target coordinates. Intelligence coordination. Mission plans.
Theater and Ballistic Missile Defense Defense-enabled freedom of maneuver.	
Position, Navigation, and Timing Global Positioning System (GPS) location information. GPS-timing enables communications coordination.	Precision target location. GPS-guided munitions.
Space Control Ensure access to national space assets and provide protection from detection by hostile assets.	Information operations via space links enabled.

Unique Navy requirements generally focus on supporting persistent operation over the oceans, and Navy input is thus centered on assuring that NSS ISR systems will meet the Navy's open-ocean needs. New space-based ISR capabilities show promise for meeting Navy needs and are currently under development. These include space-based radar (SBR), the Future Imagery Architecture (FIA), and hyperspectral imaging. Hyperspectral imaging in particular is promising in its potential to assist in separating targets from background and camouflage, especially in the ocean and littoral areas unique to Navy and Marine Corps activity.

Naval Fire Support	Maneuver	Strategic Deterrence
Blue force tracking (BFT) and red force location deconfliction. BFT integrated into common operating picture and fire support systems.	Mission preparation. Minefield monitoring. High-resolution-imagery-derived assault planning.	Monitoring production and storage of strategic threats. Imagery enabling tracking of potential threats.
Steaming direction for carrier aircraft launch. Cue appropriate ISR assets depending upon weather.	Mission planning. Amphibious assault planning requires littoral oceanographic information.	Weather cueing of ISR assets to enable continuous monitoring.
Blue force tracking. Transmission of fires requests. Creation and use of a common operating picture enabled.	Mission coordination. Creation and use of a common operating picture enabled.	BLOS detection and monitoring of threats enabled.
		Augmentation of theater defense to protect against first strike.
GPS-enabled BFT. Precision target location. GPS-guided munitions.	GPS-guided assault planning. GPS-guided vehicles.	Precision target location.
Protection against adversary detection of blue forces.	Freedom of maneuver without detection enabled.	Protection against threat of first strike against space assets.

Time-sensitive strike operations have significant dependency on high-availability, high-assurance, low-probability-of-intercept (LPI), medium-bandwidth, low-latency, all-weather communications between ISR sensors; on distributed analysts, exploitation, and fusion processors and centers; and on the diverse command-and-control systems linking strike weapons, platforms, and infrastructure. In addition, constraints on satellite communications, antenna size, portability, and ease of setup and disassembly are critical for mobile units in the field. As described in recent after-action reports on Operation Iraqi Freedom, the most reliable BLOS communications capabilities were space-based systems.[13] Improvements in such a reliable and mobile communications capability, allowing for increased numbers of supported users, were also recently identified by the Commander, Central Command, as his top priority.[14]

METOC support to Sea Strike generally refers to providing atmospheric weather forecasts to get planes off carrier decks safely and to enable appropriate mission planning and ordnance use (for example, forecasts of regional dust might affect the potential use of optically guided munitions). Weather information is primarily derived through the National Oceanic and Atmospheric Administration (NOAA) and is based on a combination of national (Geostationary Operational Environmental Satellite (GOES) and Polar Operational Environmental Satellite (POES)) and military (Defense Meteorological Satellite Program (DMSP)) environmental satellites. The needed data sets often rely on 1 km resolution data derived from the sources referred to above and augmented by commercial multispectral imagers. For full global coverage, the Navy relies on the European Space Agency's Meteorological Satellite (Meteosat), India's Indian National Satellite (INSAT), and Japan's Geostationary Meteorological Satellite (GMS). Timeliness of the data from commercial and international partners can be a tactical issue, however, since commercial data are typically delivered in days or weeks.

Having collected raw-satellite-derived data as described above, the Fleet Numerical Meteorology and Oceanography Center (FNMOC) turns out products that respond directly to the needs of commanders and planners. In particular, FNMOC prepares a nowcast or forecast virtually everywhere as a meteorological objective analysis, not just where the meteorological observations are made. FNMOC also collates, interprets, and delivers in a user-friendly form the information that commanders require in order to achieve their mission objectives, offensive or defensive.

METOC products designed to enhance warfighting capabilities in the littoral regions need finer spatial resolution and greater repeat coverage than are required for the forecasting services described above. Examples of such tactical decision aids include acoustic propagation predictions and ocean-surface, front, and eddy thermal analysis to support antisubmarine warfare operations.

Sea Shield

Sea Shield is the overarching concept describing how 21st-century naval forces will project defense over water and over land to protect and enable sus-

---

[13]Marine Corps Combat Development Command. 2003. *Field Report Marine Corps Systems Command Liaison Team, Central Iraq (April 20-25, 2003)*, Quantico, Va., May.

[14]Written response by General John Abizaid, USA, Commander, Central Command, during his congressional confirmation hearing, June 24, 2003. Available at <http://armed-services.senate.gov/statemnt/2003/June/Abizaid.pdf>. Accessed May 17, 2004.

tained influence in joint campaigns. These naval forces will protect against adversaries through the application of four critical capability areas: force protection, surface warfare, undersea warfare, and theater air and missile defense. "Sea Shield will provide a layered defense to protect the homeland, sustain access to contested littorals, and project a defensive umbrella over coalition partners and joint forces ashore in distant theaters."[15]

Sea Shield envisions naval forces that will assure their own survival (afloat or ashore) and the survival of their associated air and surface logistic forces. Sea Shield also requires that naval forces establish and maintain air control against hostile aircraft and be capable of mounting a successful defense against cruise and ballistic missile attack, both in naval operating areas and as far inland as practicable. Thus, based on the Sea Shield vision, the Navy has identified, in these four capability areas, 13 specific capabilities that contribute critical naval needs for enabling future defensive operations (see Table 2.3). Most of these capabilities are historic priorities for the naval forces, while newly emerging areas include sea-based national missile defense. Not clearly articulated in these capabilities (but implicitly embedded in the implementation of them) are the needs for denial, degradation, deception, and exploitation of enemy ISR sensors, communications, and databases; and for surveillance of the open seas to identify and track hostile shipping engaged in the transport of missiles and/or weapons of mass destruction that may present a threat to U.S. homeland security. Denial, degradation, deception, and exploitation have historically been considered a theater or national responsibility, while open-sea surveillance, to be performed in conjunction with the U.S. Coast Guard, is a new naval responsibility that will clearly require extensive space-based assets and operational coverage and analyses.

To maintain littoral superiority for naval and joint force components, communications resources must be able to support protection against unconventional threats (i.e., chemical, biological, radiological, nuclear, and environmental threats from special operations and terrorist forces) as well as threats that might be mounted by more conventional enemy ground forces. Information from space-, ground-, and sea-based and airborne ISR resources must be communicated to neutralize near-horizon and over-the-horizon threats, as well as to enable deep-ocean and littoral operations. This information is then used to support self-defense against, or neutralization of, undersea threats (including submarines, mines, submerged barriers, and obstacles) and to provide defense over land and over water against theater air and ballistic missile threats. Additional areas in which space systems specifically enable the Sea Shield capability areas are summarized in Table 2.4.

---

[15]VADM Michael Bucchi, USN; and VADM Michael Mullen, USN. 2002. "Sea Power 21 Part II; Sea Shield: Projecting Global Defensive Assurance," *U.S. Naval Institute Proceedings*, Vol. 128, No. 11, p. 56.

TABLE 2.3 Sea Shield Capabilities and Capability Areas

Force Protection	Surface Warfare	Under Sea Warfare	Theater Air and Missile Defense
Protect against Special Operations Force and terrorist threats. Mitigate effects of chemical, biological, radiological, nuclear, and environmental threats.	Neutralize near-horizon surface threats. Neutralize over-the-horizon surface threats.	Provide self-defense against subsurface threats. Neutralize submarine threats in the littorals. Neutralize open-ocean submarine threats. Neutralize mines in shallow to deep water. Breach surf zone, minefields, obstacles, and barriers.	Provide self-defense against air and missile threats. Provide maritime air and missile defense. Provide overland air and missile defense. Maintain sea-based national missile defense.

SOURCE: RADM K.J. Cosgriff, USN, Director, Warfare Integration and Assessment, "Future Force Development," slide 8, presentation to the Naval Studies Board, Washington, D.C., November 12, 2003.

Critical support for the effective utilization of all available ISR resources will depend on a combination of terrestrial line-of-sight communications for battle group defensive capability, such as that of Cooperative Engagement Capability (CEC), and space-based communications for general support in detecting, identifying, and neutralizing over-the-horizon (OTH) threats and deep-water mine fields and for linking and communications between theater and national sensor systems and command structures. Expanding the CEC to include OTH links and the introduction of other new capabilities, such as wide-area distributed undersea operations, will require the development of new, high-bandwidth, low-latency, high-assurance communications. Space-based communications will be an essential link in establishing this extended capability.

Today, most operations rely largely on theater assets to provide the necessary ISR information to effectively support Sea Shield operations. In the future, NSS ISR will be able to contribute much more to many of these operations. In particular, SBR represents a major ISR initiative to field both GMTI and SAR imagery capabilities. The scope of this initiative is such that it is very unlikely that any other space-based radar initiatives will be implemented in the next several decades. It is imperative, therefore, that the Navy understand its needs for this extended time period and ensure that these needs are articulated to and met by the SBR initiative. For example, GMTI must contend with Doppler-induced background clutter that has significantly different character for land use and marine use, signifying a potential area in which maritime use could be traded away in favor of land use.

One of the potential applications of SBR and a needed capability for Sea Shield is persistent tracking of ships at sea. Current antisubmarine warfare (ASW) and antimine operations lack persistence in making observations of offensive maritime operations by an enemy. While it is possible to observe enemy submarines at shallow depth from space and to observe from space the laying of mine fields or the navigation by the enemy through fields it has laid, the long revisit time for current NSS systems does not well support the near-continuous observations necessary. SBR is capable of enabling these missions through persistent (nearly continuous) monitoring of maritime traffic, provided the Navy can ensure that these capabilities are protected in the developing SBR program.

Space communications will be vital in linking information from BLOS sensing systems in many theater or strategic missile defense scenarios, especially by enabling geolocation and computation of the vector of inbound ballistic and cruise missile threats.

An area of Sea Shield for which no significant new NSS active ISR is currently contemplated is theater air and missile defense. An NSS capability to provide persistent AMTI radar similar to that provided by current E-2C (Hawkeye aircraft) and the Airborne Warning and Control System (AWACS) radars would dramatically expand—to global dimensions—the range for detecting and targeting airborne threats beyond the ranges currently provided by UAVs or piloted aircraft (hundreds of square miles). A specific NSS program to develop and field such an AMTI capability is not now contemplated, and creating such a system would represent a stretch with current NSS sensor technology. However, the Navy would benefit from continued S&T aimed at identifying future NSS ISR potential in this area.

A nearer-term NSS opportunity to address the missile defense mission could involve a persistent, multiple-look-angle NSS IR detection capability. The individual satellite sensors might each be similar in capability to that available with the Space-Based Infrared System-High (SBIRS-H) satellites under development. Linking of multiple satellites with a view of the warfare theater could then permit precision geolocation, detection, and tracking of theater ballistic missiles by SBIRS-H or current Defense Satellite Program satellites.

Sea Basing

Sea Basing provides the operational platform capability from which the Navy will participate in the planning and coordination of operations using FORCEnet and from which it will provide offensive strike (i.e., Sea Strike) and defensive protection (i.e., Sea Shield), as well as supporting the Marine Corps and joint forces as appropriate to their missions. The creation of a sea base is accomplished by use of the fleet assets with platform, logistics, and communica-

TABLE 2.4 Space Capabilities, Derived from Space Mission Areas, to Enable Sea Shield Capability Areas

	Sea Shield Areas
Space Mission Areas	Force Protection
Intelligence, Surveillance, and Reconnaissance (ISR) Responsive persistent ISR (imagery and signals) of static and moving targets.	Global ship tracking/maritime moving target indication. Mine (field) detection and monitoring. ISR through camouflage needed.
Meteorology and Oceanography Continuous tactical weather prediction. Acoustic/thermal modeling of the littorals.	Tactical decision aids. Weather-prediction-enabled maneuver.
Theater and Ballistic Missile Defense Defend against missile threats. Standard Missile (SM)-2 guidance at beyond-line-of-sight (BLOS) range.	All-weather, continuous ability to detect missile launches. Overland cruise missile defense mission needs airborne moving target indication.
Communications All information must be transferred, often at BLOS ranges. Sensor-analysis-decisions-shooter-weapon links.	Mission coordination. Sensor cueing and sensor netting. BLOS targeting sensor-to-shooter.
Position, Navigation, and Timing Global Positioning System (GPS) location information. GPS timing enables communications coordination.	GPS-enabled blue force tracking. Precision target location. GPS-guided munitions.
Space Control Ensure access to national space assets and provide protection from detection by hostile assets.	Protection of forces against observation by hostile ISR.

tions improvements, but without extensive use of existing port facilities or logistics ashore. "Sea Basing is the core of 'Sea Power 21.' It is about placing at sea—to a greater extent than ever before—capabilities critical to joint and coalition operational success: offensive and defensive firepower, maneuver forces, command and control, and logistics."[16]

[16]VADM Charles W. Moore, Jr., USN; and Lt Gen Edward Hanlon, Jr., USMC. 2003. "Sea Power 21 Part IV; Sea Basing: Operational Independence for a New Century," *U.S. Naval Institute Proceedings*, Vol. 129, No. 1, p. 80.

Surface Warfare	Under Sea Warfare	Theater Air and Missile Defense
Global ship tracking/ maritime moving target indication. Cruise missile detection/ cueing. Wake detection.	Adversary port monitoring. Wake detection.	All-weather, continuous overhead exo- and endo-atmospheric cruise and ballistic missile sensing needed. Airborne moving target indication desired.
Tactical decision aids. Weather-prediction-enabled maneuver.	Subsurface acoustic prediction. Undersea weather. Bathymetry. Littoral currents/tides.	Weather-cueing ISR for continuous monitoring.
All-weather, continuous ability to detect missile launches.	All-weather, continuous ability to detect missile launches.	Missions entirely overlap.
Mission coordination. BLOS communications.	Mission coordination. BLOS communications. Read-out of deployed antisubmarine warfare sensor nets.	BLOS sensor cueing and netting to enable BLOS threat missile interception.
Precision target location. GPS-guided munitions.	Precision target location. GPS-guided munitions. Unmanned underwater vehicle navigation enabled by GPS.	Precision target location.
Tactical space sensor launch enabled.		Protection against threat of first strike against space assets.

Conceptually, light combat forces would be inserted inland and their logistics requirements would be sustained by versatile air- or sea-lift delivery. Major firepower support would come from sea-launched missiles and sea-based strike aircraft. Logistic support would require the availability of large surface ships with selective off-load capabilities that could respond rapidly to the logistics needs of engaged forces. Although the design of such a new class of logistics ships has not been decided upon, preliminary studies indicate that large (100,000 to 600,000 ton displacement) vessels could be required. These vessels would need to be capable of being off-loaded using both heavy-lift lighterage and heavy-lift aircraft. The concept of sea basing is dependent on an ability to assure the survivability of the logistics base at sea and logistics forces in transit, as well as being dependent on the system response time to the needs of engaged combat elements for both logistical and fire support.

The Sea Basing concept presents a major resource management challenge: complex air and surface logistics must be managed, threats to personnel and supplies in transit between the sea base and the engaged ground combat forces must be countered or avoided, resupply of the sea base from commercial shipping must be provided for, and containerized loads must be identified and located. To structure these needs, the Sea Basing concept is built around 13 specific capabilities in the following areas: deploy and employ, provide integrated joint logistics, and pre-position joint assets afloat (as listed in Table 2.5). Note that Sea Basing implies not only the provision of logistics support at sea and inland, but it also encompasses the support of sea-based joint command and control, thus enabling a truly sea-based Joint Forces Command Center. The resulting space mission area needs and capabilities for enabling Sea Basing are listed in Table 2.6.

Sea Basing's METOC needs include provision of the weather and oceanographic predictions necessary to enable optimum track ship routing (OTSR) modeling services as well as provision of the environmental information to enable management of the lighterage component of the delivery system. Management of this component requires predictions to enable ship-to-ship cargo transfer and littoral oceanography to enable ship-to-shore cargo transfer without access to developed ports.

Of the Sea Basing concept's space-dependent needs, the one that places perhaps the most stress on current space technology is the proposed establish-

TABLE 2.5 Sea Basing Capabilities and Capability Areas

Deploy and Employ	Provide Integrated Joint Logistics	Pre-Position Joint Assets Afloat
Close the force and maintain mobility. Provide at-sea arrival and assembly. Allow selective offload. Reconstitute and regenerate at sea.	Provide sustainment for operations at sea. Provide sustainment for operations ashore. Provide focused logistics. Provide shipboard and mobile maintenance. Provide force medical services. Provide advanced base support.	Integrate and support joint personnel and equipment. Provide afloat command and control physical infrastructure. Provide afloat forward staging base capability for joint operations.

SOURCE: RADM K.J. Cosgriff, USN, Director, Warfare Integration and Assessment, "Future Force Development," slide 10, presentation to the Naval Studies Board, Washington, D.C., November 12, 2003.

ment of a forward-deployed, sea-based joint forces command center. The need is to provide the networking connectivity and communications bandwidth necessary to duplicate the capabilities of a land-based joint forces command center. On the sea base, large-scale planning will require access to large, worldwide databases, and (since most data are collected off the sea base) rapid, agile planning on the sea base will require very wide bandwidth communications to collect and assess planning information, to implement high-quality video conferencing, and to access near-real-time sensor information. An additional constraint on bandwidth is due to the DOD's current information dissemination concept of tasking, posting, processing, and using (TPPU), under which all raw sensor data are made available for all potential users.

All of these requirements bespeak a communications capability similar to that of a major base command. During Operation Iraqi Freedom, estimates of peak communications needed to support operations of the joint forces were over 750 Mb/s.[17] Since communications will often be at BLOS ranges, the requirements of bandwidth and connectivity dictate that the backbone of the needed communications links be space-based. Current Navy capabilities for space-derived communications are approximately 8 Mb/s to large-deck ships, with plans to increase this to approximately 25 Mb/s by 2009.[18] Owing to the ongoing explosive growth in space communications needs, such bandwidth plans seem at odds with the communications needs to enable Sea Basing.

FORCEnet

FORCEnet is essentially the networked communications and information collection, fusion, and processing capability required to implement the command, control, and communications functions of Sea Power 21. Although the total FORCEnet architecture has not been established, its desirable attributes and capabilities are generally understood and accepted by the Naval Services. FORCEnet focuses on the gathering, processing, transportation, and presentation of information in support of the entire scope of the Sea Power 21 vision, serving as an integrator and enabler for the three pillars Sea Strike, Sea Shield, and Sea Basing. FORCEnet relies on and builds from larger DOD initiatives, especially the Global Information Grid (GIG) and Transformational Communications (TC) concepts that implement information transmission worldwide. The DOD envisions that data from national archival sources, command messages, responses to

[17]Lt Gen T. Michael Moseley, USAF, Commander, Central Air Forces. 2003. *Operation Iraqi Freedom—By the Numbers*, Shaw Air Force Base, S.C., April 30, p. 12.

[18]CAPT John Yurchak, USN, Naval Network Warfare Command. 2003. "C5I [command, control, communications, computers, combat systems, and intelligence] Day—Progress Report Fleet Satellite Network Communications," presentation to ADM Robert J. Natter, USN, Commander, Fleet Forces Command, Norfolk, Va., March 31.

TABLE 2.6 Space Capabilities, Derived from Space Mission Areas, Needed to Enable Sea Basing Capability Areas

	Sea Basing Areas
Space Mission Areas	Deploy and Employ
Intelligence, Surveillance, and Reconnaissance (ISR) Responsive persistent ISR (imagery and signals) of static and moving targets.	Global ship tracking, monitoring, and identification. Sea base threat determination.
Meteorology and Oceanography Continuous tactical weather prediction. Acoustic/thermal modeling of the littorals.	Ocean routing. Weather prediction to enable at-sea cargo transfer.
Theater and Ballistic Missile Defense (TBMD) Defend against missile threats. Standard Missile (SM)-2 guidance at beyond-line-of-sight (BLOS) range.	Ability to defend sea base against theater and ballistic missile attack. Overland cruise missile defense mission needs airborne moving target indication.
Communications All information must be transferred, often at BLOS ranges. Sensor-analysis-decisions-shooter-weapon links.	Connectivity and coordination.
Position, Navigation, and Timing Global Positioning System (GPS) location information. GPS timing enables communications coordination.	GPS-enabled navigation.
Space Control Ensure access to national space assets and provide protection from detection by hostile assets.	Protection of forces against observation by hostile ISR.

queries, and space-based or unattended ground sensors will all traverse the GIG. FORCEnet will also provide the communications infrastructure, network protection, and information assurance functions internal to the naval network, as well as an integrated common operational and tactical database accessible Navy-wide.

The FORCEnet concept demands far more than the construction of a modern communications system. FORCEnet is intended to be a network that allows the following:

- Reach-back by forward-deployed or engaged forces to archival information relative to targeting and adversarial capabilities;

Provide Integrated Joint Logistics	Pre-position Joint Assets Afloat
Global ship tracking, monitoring, and identification.	ISR needs for the sea-based joint command-and-control headquarters.
Ocean routing/logistics scheduling. Weather prediction to enable at-sea cargo transfer.	Meteorology and oceanography needs for the sea-based joint command-and-control headquarters. TBMD needs for the sea-based joint command-and-control headquarters.
Connectivity and coordination. Logistics tasking, scheduling, and coordination.	Full communications capabilities of a joint command-and-control center afloat.
GPS-enabled navigation.	GPS-enabled navigation.
	Protection of forces against observation by hostile ISR.

- Seamless and timely dissemination of newly derived ISR information to engaged forces;
- Support of a common operational picture that provides both forward and rear command echelons with common, accurate situational awareness;
- Blue force tracking (the ability to continually identify, locate, and track friendly forces); and
- Self-synchronized logistic support of engaged units in response to the automatic tracking of weapons, food and fuel, platforms, and personnel.

Force projection and defense from forward-deployed naval platforms depend on the efficient networking of naval, national, and force nodes involved in all aspects of information production, command responsibility, and control authority through communications and computing power to meet Navy objectives. Accordingly, FORCEnet is built on 19 specific capabilities in three critical areas: communications and data networks; intelligence, surveillance, and reconnaissance; and common operational and tactical picture (COTP). These capabilities are listed in Table 2.7.

As discussed in the preceding subsections on Sea Strike and Sea Shield, most strike and defensive actions will use information from multiple ISR sources to engage targets and threats, with a goal of having the actual source of any individual information element transparent to the warfighter. As the mission of the Navy grows in the 21st century and as the theater of importance expands to global dimensions, the importance of ISR information derived from NSS assets will necessarily grow. In particular, the timeliness, relevance, quality, and quantity of information will continue to determine the outcomes of naval missions. This means that the need for technology associated with space will grow, and the technology associated with the processing, exploitation, and fusion of NSS-derived data to produce intelligence will continue to increase in importance. Table 2.8 lists the general space capabilities that will be needed to enable the FORCEnet capability areas.

TABLE 2.7 FORCEnet Capabilities and Capability Areas

Communications and Data Networks	Intelligence, Surveillance, and Reconnaissance	Common Operational and Tactical Picture
Provide communications infrastructure. Provide network protection. Provide network synchronization. Provide information transfer.	Conduct sensor management and information processing. Detect and identify targets: Fixed land targets, Moving land targets, Air and missile targets, Surface targets, Submarine targets, Mines. Provide cueing and targeting information. Assess engagement results.	Provide mission planning. Provide battle management synchronization. Provide common position, navigation, and timing and environmental information. Integrate and distribute sensor information. Track and facilitate engagement of time-sensitive targets. Track and facilitate engagement of non-time-sensitive targets.

SOURCE: RADM K.J. Cosgriff, USN, Director, Warfare Integration and Assessment, "Future Force Development," slide 11, presentation to the Naval Studies Board, Washington, D.C., November 12, 2003.

The dependence of FORCEnet on space-based communications is implicit in the FORCEnet capabilities discussed here. FORCEnet has not developed, as far as the committee can discern, a systems-engineered view of the connectivity and capability required from space-based communications systems. This is a major shortcoming that needs to be addressed so that the Navy can define and defend its requirements to the executive agents, plan and allocate resources, and articulate S&T needs.

The Defense Information Systems Agency (DISA) recently established the TPPU information dissemination concept referred to above—that is, prior to its processing and analysis by the DOD or intelligence communities, all information is to be posted and available for all potential users.[19] This concept has an unintended potential to greatly increase communications bandwidth demands. Under the TPPU concept, it is envisioned that large numbers of users will be accessing relatively large, unprocessed data sets directly collected from sensors, rather than relying on reach-back analysis capabilities that would prepare, digest, and forward relevant condensed information. For example, the large amounts of data that the F-18 E/F mounted tactical airborne reconnaissance pod system collects under TPPU will be posted for all potential users to access. The many possible ways to handle these vast amounts of data (e.g., shipboard server storage, terrestrial data warehousing, and so on) all involve very high bandwidth space-based communications between maritime warfighting platforms and storage locations. So, worldwide wide-bandwidth, high-availability space-based communications are essential for implementation of the TPPU concept.

Unfortunately, many moving naval platforms do not have the full communications capability to receive and transmit all of the data required for an implementation of the FORCEnet concept. Considerable effort has been and will continue to be invested in the development of systems that ultimately will allow all naval mobile platforms to be full participants in the naval force network. Additionally, future naval platforms are being designed to be ever more critically dependent on network connectivity and off-board sensors. In fact, recently the concepts of operations underlying the Navy's planned littoral combat ship (LCS)—a small striking and support ship to enable operations in the world's littorals—was described as "the first ship built from the keel up to operate as an element of the highly networked naval force envisioned under FORCEnet. This will allow the design of a core LCS vessel carrying only the most essential onboard sensors for self-protection and mission accomplishment."[20]

While the backbone communications system that will support FORCEnet has not been defined, there is a high probability that ultimately the laser-linked

---

[19]For more information on TPPU, see <http://ges.dod.mil/about/tppu.htm>. Accessed May 13, 2004.

[20]RADM H.G. Ulrich III, USN; and RADM Mark J. Edwards, USN. 2003. "The Next Revolution at Sea," *U.S. Naval Institute Proceedings*, Vol. 129, No. 10, p. 67.

TABLE 2.8 Space Capabilities, Derived from Space Mission Areas, Needed to Enable FORCEnet Capability Areas

	FORCEnet Areas
Space Mission Areas	Communications and Data Networks
Intelligence, Surveillance, and Reconnaissance (ISR) Responsive persistent ISR (imagery and signals) of static and moving targets.	
Meteorology and Oceanography Continuous tactical weather prediction. Acoustic/thermal modeling of the littorals.	Local weather determines communications availability and potential bandwidth.
Theater and Ballistic Missile Defense (TBMD) Defend against missile threats. Standard Missile (SM)-2 guidance at beyond-line-of-sight (BLOS) range.	
Communications All information must be transferred, often at BLOS ranges. Sensor-analysis-decisions-shooter-weapon links.	Space-based communications provide most current BLOS links.
Position, Navigation, and Timing Global Positioning System (GPS) location information. GPS timing enables communications coordination.	Timing enables communications and network coordination.
Space Control Ensure access to national space assets and provide protection from detection by hostile assets.	Space control protects forces against loss of communications capabilities. Tactical communications satellite launch to fill immediate needs.

GIG envisaged in the proposed Transformational Communications Architecture (TCA) will be employed. In brief, the GIG will employ a number of geosynchronous satellites that are capable of satellite-to-satellite and satellite-to-ground communications over wide-bandwidth laser links. Such links, however, have been found difficult to connect directly to naval vessels owing to rapid, unpredictable ship motions and the general maritime environment. One proposed solution is for each battlegroup to be linked by high-bandwidth, line-of-sight radio frequency communications to a high-altitude UAV that is then laser-linked to the TC satel-

Intelligence, Surveillance, and Reconnaissance (ISR)	Common Operational and Tactical Picture (COTP)
Space-based ISR is a major component of total ISR.	ISR provides imagery and intelligence for generation of COTP.
Weather cueing needed to schedule ISR assets. METOC data are a component of total ISR.	Weather prediction must be part of COTP. Common environmental information.
TBMD sensor data are a component of total ISR.	Provision of missile threats to the COTP. Overland cruise missile defense mission needs airborne moving target indication.
ISR information must be communicated. BLOS communications enable reach-back analysis and support.	COTP is built from coordinated data sets and must be distributed to be useful.
Many ISR resources derive location data from GPS.	GPS-derived location of all blue and red forces must insert into COTP.
Space control protects forces against loss of ISR capabilities. Tactical ISR satellite launch to fill immediate needs.	Space control enables limitation of adversary space-derived ISR information.

lites. During the ongoing development of the GIG, TCA, and FORCEnet, all such proposals need to be clearly articulated and addressed.

One of the key parameters of any tactical-information communications system architecture is latency. ISR systems in general lump together three products of space-based sensors (intelligence, surveillance, and reconnaissance) and do not include requirements needed to support time-sensitive target tracking. The desired latency of deriving ISR data may vary from milliseconds for theater ballistic missile warnings to months for the identification of potential hostile weapons production facilities. Thus, FORCEnet and its associated space links will need to enable the most stressing latencies that can be accommodated.

As discussed in the subsection above on Sea Basing, FORCEnet will also need to enable the provision of the communications, ISR, METOC, and TBMD capabilities necessary for a sea-based joint forces command center. This need presents a major shortfall in the scale of the bandwidth that the Navy can support, and the gap between ability and bandwidth needs will tend to increase as new systems, products, and capabilities become deployed.

Sea Power 21: Capability Dependence on Space Mission Areas

As with other components of the DOD, naval forces are now highly dependent on support from space for a broad range of military functions. In fact, one of the most significant findings of this committee is that the Navy's broad vision of future naval warfare operations, as expressed in Sea Power 21, cannot be executed without extensive and continuous support from space-based systems.

Many broad areas of commonality exist among the Services' increasing reliance on support from space. However, naval forces also have some very unique requirements. These are generally driven by the mobility of naval forces across broad ocean areas, their need for accurate marine environmental data, the nature of littoral military operations, and the limited antenna configurations suitable for installation on naval ships and craft. The unique requirements of naval forces are discussed throughout Chapter 4 of this report, but they can be usefully summarized as follows:

- Broad ocean and littoral surveillance of potentially hostile aircraft, missiles, and ships;
- Secure, reliable data and voice communications to and from rolling, pitching, heaving, and turning ships;
- Detailed temporal and real-time measurements of maritime environmental conditions, including littoral land areas; and
- Maritime navigation aids.

Based on the threats described above and on the ensuing Navy organization around the four Sea Power 21 components, the committee provides, in Figure 2.1, a summary assessment of Sea Power 21 capabilities mapped against the NSS space mission areas. While every one of the Sea Power 21 pillars has multiple critical dependencies on space assets, the overwhelming reliance of FORCEnet on space-based capabilities reinforces the need for sustained and effective Department of the Navy participation in the NSS community. Each of these space mission areas is currently undergoing transformation in the form of new systems or major block upgrades. The Department of the Navy does not appear to recognize or take advantage of the opportunities presented by this transformation or by

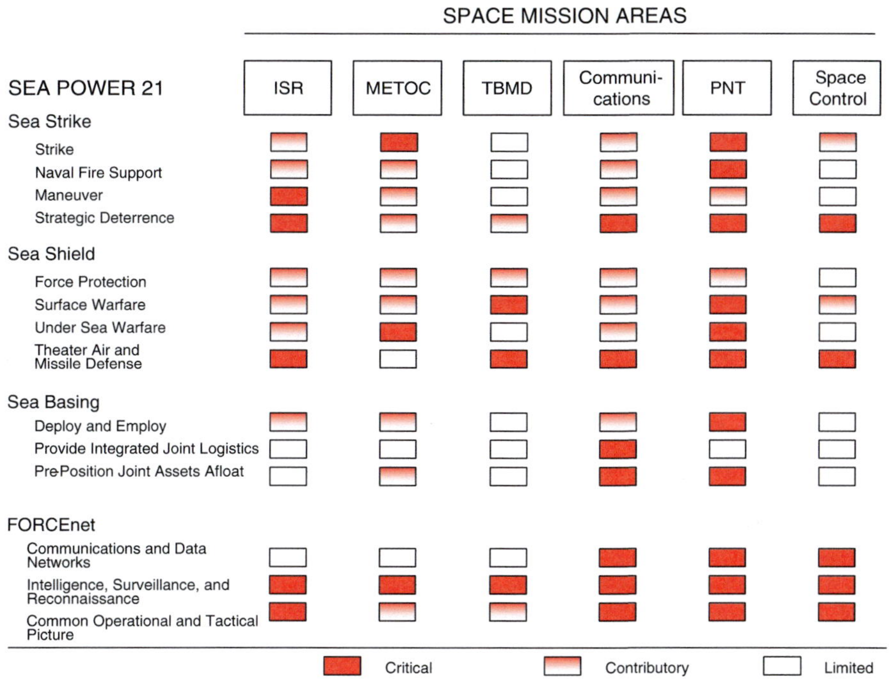

FIGURE 2.1 Dependency of Sea Power 21 on National Security Space mission areas. (A list of acronyms is provided in Appendix G.) NOTE: A "critical" dependency reflects a space mission area that is considered absolutely necessary for accomplishment of the particular Sea Power 21 capability. For example, without access to space-based ISR information the Navy could not generate the Common Operational and Tactical Picture needed to support FORCEnet; thus the block linking these two areas is denoted critical. A "contributory" dependency reflects a space mission area that will provide support for accomplishing the particular Sea Power 21 capability.

the establishment of the DOD Executive Agent for Space. Thus, the Navy does not appear to be defining and integrating the unique and pervasive capabilities of NSS architectures and systems to generate the critical components of Sea Power 21.

SPACE POLICY

Since the current Department of the Navy space policy was established in 1993,[21] significant changes have occurred in both the national security environ-

[21]John H. Dalton, Secretary of the Navy. 1993. SECNAV Instruction 5400.39B, "Department of the Navy Space Policy," Washington, D.C., August 26.

ment and the DOD's posture toward the development, support, and operation of space-reliant technologies. In particular, the recently established homeland defense mission tasks the Navy (in concert with the U.S. Northern Command, the Coast Guard, and other federal agencies) with protecting the maritime approaches to the United States.[22] The Navy's specific roles and responsibilities in support of this mission remain undefined, warranting clarification, but they will likely entail reliance on the development of new space capabilities, such as near-continuous, open-ocean surveillance of all surface craft. In addition, the Department of the Navy need only review the lessons learned from the Desert Storm experience in comparison with those arising from Operation Iraqi Freedom to be reminded of the ever more critical role that space systems are assuming in the successful prosecution of U.S. combat operations. Space is also unique for its ability to enhance everyday lives with services such as communications, GPS, and weather support. Given this growing importance of and dependence on space—for military and civil, peacetime and wartime purposes—the lack of an updated Department of the Navy space policy leads to a potential dilution of effort, resulting from a lack of naval focus on space as a needed participant in future warfighting.

The DOD, on the other hand, has made several recent changes in its approach to the support and development of space mission areas. One such change was the promulgation in 1999 of a DOD space policy, which states, "The primary DOD goal for space and space-related activities is to provide operational space force capabilities to ensure that the United States has the space power to achieve its national security objectives."[23] More importantly, in 2003 the Secretary of Defense promulgated DOD Directive 5101.2,[24] reorganizing the oversight of NSS mission areas through the creation of a single oversight director—the DOD Executive Agent for Space, who is also the Under Secretary of the Air Force and director of the National Reconnaissance Office. This allocation of NSS activities to the DOD Executive Agent for Space (and thus to the Air Force) has led many in the Navy to indicate to this committee their concern that the Air Force will not adequately protect and support the development of critical naval space needs and requirements.

In addition, through DOD Directive 5101.2, the Secretary of Defense has directed the Department of the Navy to fulfill a series of specific policy, resource, and support activities to assist the DOD Executive Agent for Space in developing naval space capabilities. The duty to draft recommendations for the Navy in

---

[22]ADM Vern Clark, USN, Chief of Naval Operations. 2004. "CNO Guidance for 2004," Department of the Navy, Washington, D.C., February.

[23]Department of Defense. 1999. DOD Directive 3100.10, "Space Policy," William Cohen, Secretary of Defense, July 9, p. 6.

[24]Department of Defense. 2003. DOD Directive 5101.2, "DOD Executive Agent for Space," Paul Wolfowitz, Deputy Secretary of Defense, June 3. The complete text of this directive is presented in Appendix B of this report.

support of this directive has been assigned to the Under Secretary of the Navy, a position that was vacant before the promulgation of the directive.[25] In the intervening time, the Navy has forwarded no documents or policy statements to the DOD Executive Agent for Space detailing how the Department of the Navy will act to support the DOD Executive Agent for Space.

Recommendation 2.1. The Secretary of the Navy should task the Chief of Naval Operations and the Commandant of the Marine Corps to formulate and take steps to establish a new Department of the Navy space policy.

This space policy should provide a framework for Department of the Navy participation in the planning, programming, and acquisition activities of the Department of Defense Executive Agent for Space, to include definition of the Navy Department's relationship to National Security Space activities. A primary objective of the Department of the Navy space policy should be to focus attention on space mission areas critical to the successful implementation of Naval Power 21 as well as on other national maritime responsibilities such as homeland defense.

---

[25]A new Under Secretary of the Navy was nominated on February 6, 2004, and confirmed on October 8, 2004, while this report was undergoing classification review.

3

Roles and Responsibilities: Meeting Naval Space Needs

Space-based support for many of the Navy's current and future operations has much in common with the support needed by other military Services. For example, targeting information derived from space assets needs to be passed to naval strike units as accurately and promptly as it is delivered to other strike forces. Similarly, jam-resistant Global Positioning System (GPS) signals are needed equally by Navy, Air Force, Army, and Marine Corps precision weapons. As a result of the long and continuing history of the Navy, its operational and technical communities embody considerable expertise both on naval needs for support from space and on the underlying scientific and technical approaches that are needed for the support to be successful. But despite the many broad areas of commonality, naval forces continue to have unique requirements. As discussed in Chapter 2, these requirements are generally driven by the mobility of naval forces across broad ocean areas, by their need for accurate marine environmental data, by the nature of littoral military operations, and by the limited antenna configurations suitable for shipboard use.[1]

Initially the Navy attempted to meet many of its unique requirements for space support with Navy-funded space programs. For instance, Transit—the first satellite-based navigation system—was developed and fielded by the Navy (in conjunction with Johns Hopkins University's Applied Physics Laboratory) to meet the Polaris missile system's needs for highly accurate positional information.[2] The Naval Research Laboratory (NRL) represents a particular focal point

[1]For further elaboration, see the section entitled "Strategic Environment" in Chapter 2.

[2]See Appendix A for further detail on the Navy's history in space.

for many of the early basic and applied technologies that proved important to naval operations.

In recent years, the management of most of the space programs that provide common support to military users has been transferred to the Air Force. Included are such programs as the Navigation Satellite Timing and Ranging (NAVSTAR)/GPS precision navigation system; the Military Strategic, Tactical, and Relay Satellite (MILSTAR) system—the Department of Defense's (DOD's) current secure communications satellite program; and the Defense Meteorological Satellite Program (DMSP). The Navy continues to fund and manage the succeeding generations of legacy ultrahigh-frequency (UHF) communications satellite programs that support not only fleet operations, but many of the needs of other military forces for tactical communications.[3]

Outside the DOD, the Navy has been a partner with the National Oceanic and Atmospheric Administration (NOAA) in most of the environmental satellite programs that are important to naval operations, and it has acted aggressively to field experimental meteorology and oceanography (METOC) platforms to test new sensing concepts. Naval officers have always held leadership positions in the National Reconnaissance Office (NRO) wide-area passive electronic intelligence (ELINT) programs focused on tracking Soviet and other ships at sea. Recently, Navy participation in NRO programs has been expanded to include all NRO mission and support areas.

The Navy continues to maintain some expertise in space through the ongoing work at NRL; formal naval involvement in NRO, Defense Advanced Research Projects Agency (DARPA), NOAA, and National Aeronautics and Space Administration (NASA) programs; less formal participation in some Air Force programs; and the development and acquisition of the only major space system that remains the Navy's responsibility—UHF communications satellites.

Thus, the Navy has established an ad hoc relationship with all relevant partners in all needed space mission areas. Not only is the Navy's response tailored across the space mission areas, but it is also tailored across the five elements for effective participation in any program: (1) requirements, (2) acquisition, (3) science and technology (S&T), (4) experimentation, and (5) personnel. Table 3.1 lists the Navy's partner agencies for each of the space mission areas mapped against the five interaction elements.

To better coordinate the current ad hoc nature of Navy participation across the space mission areas, the Navy first needs to focus its awareness on these five participating elements. Only then can effective partnering plans that are specific to the space mission area be established. The following section addresses how the Navy can participate across the entire spectrum of space mission areas, and Chapter 4 provides specific guidance tailored to each of these areas.

---

[3]These programs include the Fleet Satellite (FLTSAT) Communications System, UHF Follow-on (UFO), and the Mobile User Objective System (MUOS).

TABLE 3.1 Navy Partners for the Space Mission Areas and Interaction Elements

	Space Mission Areas					
Interaction Elements	ISR	METOC	TBMD	Communications	PNT	Space Control
Requirements	NRO, Air Force, JCIDS	NOAA	MDA, JCIDS	ASD(NII), JCIDS	Air Force GPS-JPO, JCIDS	All Services, JCIDS
Acquisition	NRO, Air Force	NOAA	MDA, NRO	Air Force	Air Force GPS-JPO	All Services
Science and Technology	NRO, DARPA	NOAA, NASA	MDA, Air Force	Air Force	Air Force GPS-JPO, Navy	DARPA
Experimentation	NRO	NOAA, NASA	MDA	Air Force	Air Force GPS-JPO	DARPA
Personnel	Executive Agent for Space, NRO	NOAA	Executive Agent for Space, MDA	Executive Agent for Space, NRO	Executive Agent for Space, Air Force	Executive Agent for Space, all Services

NOTE: A list of acronyms is provided in Appendix G.

THE DOD EXECUTIVE AGENT FOR SPACE AND NEW MANAGEMENT STRUCTURE

As discussed above, until recently, the role of each of the military Services in support of space programs has been decided case by case, rather than in consonance with all the Services. In response to this situation, in 2001 the *Report of the Commission to Assess United States National Security Space Management and Organization* recommended that all DOD space activities be assigned to the Secretary of the Air Force as DOD Executive Agent for Space.[4] The intention of this reorganization was to improve the efficiency and effectiveness of the U.S. military space programs. The recommendation was initially approved by the Secretary of Defense in a memorandum of October 2001,[5] and was formally

[4]Commission to Assess United States National Security Space Management and Organization. 2001. *Report of the Commission to Assess United States National Security Space Management and Organization*, Washington, D.C., January 11.

[5]Secretary of Defense Memorandum. 2001. "National Security Space Management and Organization," Donald Rumsfeld, Secretary of Defense, Department of Defense, Washington, D.C., October 18.

implemented on June 3, 2003, by issuance of DOD Directive 5101.2.[6] In a separate but generally consistent decision, the Secretary of Defense also made a onetime adjustment between Service out-year funding totals to account for the transfer of the Navy's space surveillance program from the Navy to the Air Force.

The decision to assign the Under Secretary of the Air Force as the DOD Executive Agent for Space appears to have posed a major dilemma for the Navy—one that, in the opinion of the committee, has yet to be satisfactorily resolved. Some in the Navy have indicated a desire to rely on the Air Force to fund and provide all of the military support from space that the Navy believes it needs, with only minimal naval management attention and no naval funding beyond continuation of the UHF programs. For the Navy just to act as a "ruthless customer," however, is not the intent of the DOD directive, which clearly tasks the Navy with continued responsibility for funding and managing a range of space-related activities that are uniquely important to naval warfighting. Nor is it realistic to expect the Air Force, in the absence of strong naval management attention or funding, to assign high priority to the specialized, space-related technical and operational expertise that may be uniquely important to supporting naval warfare.

In contrast to the Navy's apparent response within the DOD to the establishment of the DOD Executive Agent for Space, outside the DOD the Navy explicitly acknowledges its responsibilities to fund unique naval needs in such programs as NOAA's National Polar-orbiting Operational Environmental Satellite System (NPOESS). However, recent actions such as cancellation of the Naval EarthMap Observer (NEMO) and Geosynchronous Imaging Fourier Transform Spectrometer (GIFTS) programs give the appearance that the Navy is withdrawing support from the development of future satellite systems.

The responsibilities of the Secretary of the Navy as delineated in the new directive form a useful framework for assessing naval organizational and funding arrangements for meeting the Navy responsibilities, not only in its interface with the DOD Executive Agent for Space, but also with NRO and NOAA. Prior to issuance of DOD Directive 5101.2, the Secretary of the Navy had delegated all responsibilities related to space to the Under Secretary of the Navy. In the absence of a new, formal delegation of responsibility by the Secretary of the Navy, the committee presumes that this delegation remains in effect and finds it to be appropriate, given the need to closely coordinate Navy and Marine Corps activities regarding space programs. However, the committee does note that since the establishment of the DOD Executive Agent for Space, the position of Under Secretary of the Navy has remained vacant. This vacancy has led to a concern that

---

[6]DOD Directive 5101.2 assigns responsibility as Executive Agent for Space to the Secretary of the Air Force, but goes on to further delegate the responsibility to the Under Secretary of the Air Force. The complete text of DOD Directive 5101.2 is presented in Appendix B of this report.

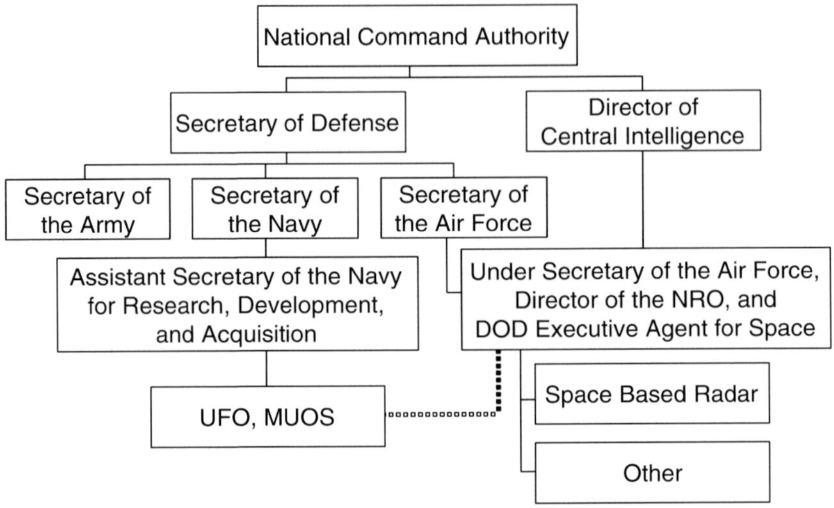

FIGURE 3.1 Command relationships within the National Command Authority to denote Navy and other Service support to National Security Space missions. NOTE: A list of acronyms is provided in Appendix G.

there is currently no naval advocate for establishing and defining the responsibilities of the Navy governed under DOD Directive 5101.2.[7]

Section 6.3 of DOD Directive 5101.2 defines specific roles and responsibilities for the heads of each of the DOD components, both with respect to their own component's internal activities and with respect to their interaction with the DOD Executive Agent for Space. The overall relationships of each of the Service Secretaries (including the Secretary of the Navy) envisioned under the new DOD Executive Agent for Space structure may be depicted as in Figure 3.1. Table 3.2 summarizes the 16 responsibilities assigned to the Secretary of the Navy. This list of responsibilities forms a useful framework for assessing the degree to which the Navy appears to be responding to the current DOD guidance in the area of space activities, which in turn provides some insight into the areas where more naval attention may be needed.

While the committee did not evaluate in detail the quality of the Navy's work in each of the 16 areas listed in Table 3.2, the table does include the committee's

[7]One positive step toward coordinating Navy space activities was reported in mid-April 2004; the article indicates that the Navy will create a Program Executive Office for Space: Amy Butler. 2004. "Navy to Establish Space Program Executive Officer," *Defense Daily*, April 14, p. 5.

TABLE 3.2 Responsibilities for Navy Support of National Security Space Programs and Committee Perspective on the Navy's Current Effectiveness in Meeting the Stated Responsibilities

Responsibility		Current Effectiveness[a]
6.3	The Secretary of the Navy shall:	
6.3.1	Participate in the planning, programming, and acquisition activities of Department of Defense (DOD) Executive Agent for Space. In support of these activities provide to DOD Executive Agent for Space information regarding:	
6.3.1.1	Space science and technology priorities, programs, and funding.	Red
6.3.1.2	Operational requirements for space and space-related systems.	Red
6.3.1.3	Strategies, plans for space systems, force structure, capabilities, measures of performance, and schedules.	Red
6.3.1.4	Approved programmatic and budget data for space programs.	Green
6.3.1.5	Space acquisition program data.	Green
6.3.1.6	Key indicators reflecting the status of, or changes to naval space cadre.	Yellow
6.3.1.7	Readiness of space forces when there are implications that may assist DOD Executive Agent for Space in addressing a space system deficiency.	Yellow
6.3.1.8	Recommendations on priorities for space support to the naval warfighter.	Red
6.3.1.9	Operations concepts for space and space-related systems.	Red
6.3.2	Provide space strategies, plans, and program information to DOD Executive Agent for Space for review, coordination, and integration into the National Security Space program and to support DOD-wide planning, programming, and acquisition.	Red
6.3.3	Submit space needs and requirements to DOD Executive Agent for Space for integration into space plans and major space program requirements documents as well as associated acquisition programs prior to submitting requirements to the Joint Requirements Oversight Council. Where possible, resolve issues with DOD Executive Agent for Space.	Red
6.3.4	Develop Department of the Navy requirements and concepts for: space systems; space doctrine, education, and training requirements and standards; space research, development, testing, evaluation, and acquisition; related military construction; and space-related strategy and operations. Provide such information to DOD Executive Agent for Space.	Red

continues

TABLE 3.2 Continued

Responsibility	Current Effectiveness[a]
6.3.5 Develop and maintain a sufficient cadre of space-qualified personnel to support the Department of the Navy in space planning, programming, acquisition, and operations. Support DOD Executive Agent for Space with space cadre personnel to represent the Department of the Navy in DOD-wide planning, programming, and acquisition activities.	Yellow
6.3.6 Recommend space and space-related planning and programming guidance for Department of the Navy programs to the Under Secretary of Defense for Policy and the Director of Program Analysis and Evaluation for consideration in their formulation of planning and programming guidance documents. Inform the Under Secretary of the Air Force, the Under Secretary of Defense for Acquisition, Technology and Logistics, the Under Secretary of Defense for Intelligence, and the Assistant Secretary of Defense for Networks and Information Integration of such submissions.	Yellow
6.3.7 Continue to develop, acquire, and fund space research, development, and acquisition programs that meet Department of the Navy requirements and submit such program information to DOD Executive Agent for Space.	Red
6.3.8 Advise DOD Executive Agent for Space on Program Objective Memorandums that significantly change any program subject to review during the program assessment for space, before submission to the Office of the Secretary of Defense.	Yellow

[a]Effectiveness evaluation: Green—adequately funded and staffed and largely effective; Yellow—marginal situation that warrants more attention; Red—little or no evidence of responsive naval management actions.
SOURCE: See Appendix B in this report for the full text of DOD Directive 5101.2, from which this table is derived.

subjective evaluation of the relative attentiveness of the Navy's response to each listed responsibility. The right-hand column summarizes the committee view of how effective the Navy currently is in fulfilling each of these responsibilities: "Green" indicates that an activity is judged to be adequately funded and staffed and largely effective; "Yellow" indicates a marginal situation that warrants more attention; and "Red" signifies that the committee saw little or no evidence of naval management actions that could be construed as responsive, suggesting the need for more management attention or resources or both. In some of the areas of responsibility judged to be in the last (Red) category, ad hoc Navy support has been (or is) provided by the Navy for single programs, without broad and enduring support arrangements having been established. More detailed rationale for the evaluations is contained in the section below on "Navy Space Support."

Often missing from the ongoing Department of the Navy discussions on space are Marine Corps participation and activities supporting space. Historically, the Marine Corps has relied on the Navy and other Services for the provision of all of its space capabilities, with little direct involvement in space capabilities development. In fact, currently the Marine Corps does not operate any space assets. Operationally, however, the Marine Corps has established itself as a prime user and developer of the concepts of operations, tactics, training, and procedures necessary to utilize space assets and capabilities.

The Marine Corps recently expanded its space-related activities in an effort to become more involved with the development of space systems. These activities include the establishment of a Marine Corps space cadre, a Marine Corps space working group to foster dialogue across the Marine Corps on space issues, and a Marine Forces component command attached to the U.S. Strategic Command. Members of the Marine Corps space cadre are being assigned to these new organizations and are also being assigned throughout the NSS structure. This new level of space-related activity appears to be resulting in the Marine Corps's becoming recognized as a stakeholder in space matters; it is also perceived that the Marine Corps is moving toward integrating space systems development into the development of its operational capabilities.

NAVY SPACE SUPPORT

The Navy's needs in space can be satisfied by focusing on the support elements: requirements, acquisition, science and technology, experimentation, and personnel. Inspection of Table 3.2 shows that these elements also summarize the responsibilities cited by DOD Directive 5101.2. Hence the discussion below expands on these elements.

Requirements Generation and Concepts of Development

The committee finds that the Navy currently lacks a rigorous operational analysis process for identifying the detailed and cost-effective requirements and priorities for space support needed to accomplish the goals of Sea Power 21. Coordination was found to be either deficient or nonexistent among the various naval offices with responsibilities for developing operational concepts that include support from space. As a result, there is no strategic plan for the Navy's role in space, and there are no clearly articulated needs for naval support from space programs that can guide the allocation of resources by naval leadership. Particularly troublesome is the lack of thorough operational analysis of the likely need for greatly increased communications throughput to and from naval forces at sea. A rigorous analytic process, including appropriate modeling and simulation, would permit the identification and evaluation of an appropriate range of space-based and non-space-based methods of providing support. The resulting cost-benefit assessments would serve the Navy well in making hard choices as to the allocation of resources.

Such assessments start with potential alternate concepts of operations. These concepts of operations are postulated by the operating force commanders for the assessment of near-term needs, and by an appropriate group charged with the responsibility of formulating potential long-term, future joint operational concepts in full consideration of the technology that could become available.

The committee recognizes that as the concepts of operations are refined, the resulting naval space-based needs must be coordinated with the overall requirements-generation processes, which for space-related issues must now be coordinated with the U.S. Strategic Command. Currently, operational inputs for the Department of the Navy's space-related requirements are collected and prioritized by the Naval Network Warfare Command (NETWARCOM), with support from the Naval Network and Space Operations Command. NETWARCOM then passes these priorities to the Office of the Chief of Naval Operations, via the Commander, Fleet Forces Command, for further prioritization and development. Space-related requirements of the Marine Corps are developed through the Information Operations and Space Integration Branch of the Strategy and Plans Division of Headquarters, U.S. Marine Corps, in coordination with NETWARCOM and the Marine Corps Strategic Command. The collected naval requirements are then supported through the joint requirements process (i.e., the Joint Capabilities Integration and Development System, or JCIDS.[8] The committee noted very little evidence of coordination or consultation among these offices on concepts of operations, requirements generation, or needs prioritization involving space support. In addition, most of the commands and processes referred to above have recently been reorganized, thus creating additional confusion.

One particular concern is the lack of a concept of operations and supporting analysis that could help establish reasonable goals for the communications bandwidth requirements to and from Navy ships. Careful analysis of potential bandwidth needs is particularly important as the DOD Transformational Communications Architecture (TCA) and Global Information Grid (GIG) are planned and developed. It appears that without a major change in this area, future naval ships are likely to deploy with a communications capacity that could be as much as two orders of magnitude below what will likely be needed for full participation in network-centric operations.[9] The result could precipitate repetitions of the situation in Operation Desert Storm when the Navy's participation was constrained by

[8]The JCIDS process is based on top-level strategic direction, provided by the Joint Requirements Oversight Council (JROC), to guide development of new capabilities. Capability recommendations and requirements are developed by the Services and evaluated by the JROC in consideration of how to optimize joint force capabilities and maximize interoperability. All major new programs are expected to participate in the JCIDS process.

[9]See the section entitled "Space-Based Communications" in Chapter 4 for further details.

the inability of the battle groups to receive complex air tasking orders electronically. Long lead times are involved in this area because of the need for appropriate antenna configurations to be included early in the design of new ships and combat aircraft; hence the importance of early and comprehensive needs analysis.

Unfortunately, the committee saw little evidence of a broad systematic approach to concepts development and requirements generation based on thorough cost-benefit analysis that can support naval forces in general or establish the appropriate role for space support in particular.[10]

Recommendation 3.1. The Chief of Naval Operations (CNO) should task the appropriate organizations—including the Commander, Fleet Forces Command; the Deputy CNO for Warfare Requirements and Programs; and the Deputy CNO for Resources, Requirements, and Assessments—to strengthen the Navy's requirements process for identifying space capability needs.

Specifically, the Navy should increase its support of operations research, systems analysis, and systems engineering (both internally and externally performed), since the Navy appears to lack sufficient resources in these areas. Operational analysis is central to the process of integrating needs across Sea Power 21 capability areas and National Security Space mission areas. The results of this analysis should be articulated for purposes of prioritization to the appropriate organizations—those with responsibility for requirements, acquisition, science and technology, and experimentation. In this process, these organizations should use common simulation, modeling, and analysis tools that are also compatible with the Joint Capabilities Integration and Development System.

Acquisition

Although Navy participation in NRO acquisition programs has been appropriate and effective, the Navy has fallen short in meeting its responsibility to participate in important DOD space acquisition programs, including Space Based Radar (SBR) and TCA. It is of particular concern that many of the DOD space programs on which naval forces will depend in the future are critical to the effectiveness of such forces, and these programs are so technically complex—with myriad potential internal cost and performance trade-offs—that a hands-off approach which defers excessively to the judgment of the DOD Executive Agent for Space risks the fielding of systems that will be inadequately responsive to the Navy's needs. Furthermore, there is concern that, without direct naval participation in the early development activities important to future space support, the

---

[10]For additional information on operations research and analysis, see the Military Operations Research Society Web site at <http://www.mors.org>. Accessed May 12, 2004.

Navy will be poorly equipped to meet the specific responsibilities relating to acquisition management outlined under the new DOD Executive Agent for Space structure.

A hands-off approach would rely on expectations that the performance of future major DOD space programs will meet the Navy's needs even though the Navy only provides early capabilities requirements. Such expectations are fraught with problems. When faced with otherwise-unaffordable cost growth, managers of large, complex acquisition programs are typically empowered to trade away performance features that they deem to be less important in favor of those that they deem more important. Unless the Navy remains an active participant during the cost management stage of such programs, important naval needs may well go unmet.

For example, in the SBR program—a major new effort being established under the DOD Executive Agent for Space charter—the committee's perception is that important maritime needs have to date been neglected or not properly prioritized.[11] This lack of attention to naval space needs derives, in part, from the lack of up-front cost-benefit analysis of the utility of the SBR concept to Navy needs, the lack of naval funding contributions, the lack of a robust naval presence in the SBR program office, and the related lack of management attention by senior naval leadership. As a result, the Navy appears to be struggling to keep up, resorting to last-minute nonconcurrence in decisions of the Joint Requirements Oversight Council (JROC) when programs do not properly reflect naval requirements. Similar concerns exist with respect to the Navy's participation in the TCA program. In part, such lack of interaction may be the result of an apparent failure of Navy operations analysis to show that incremental funds going to space-related needs are worthwhile in competition with other expenditures.

In contrast, the Navy has a strong record of active, effective partnership in the concept development, requirements generation, experimentation, acquisition, and operation of space systems under the purview of the NRO. The Navy provides the leadership of some of these programs, as well as supplying appropriate numbers of highly qualified professional staff to other NRO program and support areas. The Navy also has a strong record of partnership with NOAA, including a Navy-staffed office at NOAA.

While the Navy is not directly supporting many future and developing programs, it has consistently supported initiatives (through the Navy's Technical Exploitation of National Capabilities Program (Navy-TENCAP)) aimed at fulfilling Navy needs through the exploitation of existing (on-orbit) space assets. TENCAP programs exist for all four Services and, by congressional mandate,

---

[11]One such issue that ground moving target indication (GMTI) must contend with is a Doppler-induced background clutter that has significantly different character for land use and marine use; hence, the optimization of SBR for land use may negatively impact its marine use.

must limit their activities to the exploitation of existing National Security Space (NSS) systems. Even within this mandate, however, the Navy-TENCAP efforts have made real gains in system performance and have been able to achieve new naval capabilities. Given the strength and demonstrated utility of the Navy-TENCAP effort, the committee notes that the Navy has not sought to create TENCAP-like efforts aimed at experimentation and exploitation of space systems that are still under development (such as the Future Imagery Architecture (FIA)). Such an up-front analytic capability could serve the Navy well in understanding the utility and limitations of developing space systems.

It appears that the Navy is reluctant at present to adequately fund either internal naval space activities, such as a Navy space science and technology base, or elements of DOD space programs that are not directly managed by the Navy but that could be important to naval operations. Current responsibilities consistent with the DOD Executive Agent for Space structure require the Navy to program and budget not only for those space programs for which the Navy has been assigned the lead (for example, the MUOS program), but also for the marginal costs of other DOD space programs that meet unique naval needs. Examples of such potential, unique naval needs include broad ocean surveillance capabilities for SBR and specialized seagoing requirements for TCA. In addition, to the extent that space programs include a need for adequate terrestrial and shipboard communications terminals and antennae, these needs also warrant adequate resourcing.

In general, the earlier that a Navy need can be prioritized and its utility clearly identified, the easier that need becomes to support through the requirements generation and acquisition process. The Navy will be much more successful in assuring that its needs are met if it can contribute the needed up-front involvement and provide incremental funding. In this regard, the committee recognizes that such arrangements require careful management attention in order to avoid the impression that a contributing Service is being taken advantage of. Nevertheless, experience shows that such arrangements can work well when carefully overseen.

Recommendation 3.2. The Secretary of the Navy should ensure that the naval forces are adequately staffed and supported to influence National Security Space programs that have the potential to meet important naval space needs.

The Navy should engage early, with sufficient technical and management depth to influence the requirements generation, resourcing, and acquisition of new space systems being developed by the Air Force, National Oceanic and Atmospheric Administration, National Reconnaissance Office, and other (commercial and government) partners. In addition, the Navy's engagement should include tasking appropriate naval commands to provide inputs for and participate directly in NSS activities (see Chapter 4).

Science and Technology

While the Navy has a long history of directly supporting scientific development, experimentation, and testing of important new space capabilities (as detailed in Appendix A), at present it no longer acts like a full-fledged member of the national space S&T community. Navy funding and management of space S&T appear to be inadequate to support the Navy's needs both for broad expertise in space science and for the particular technical expertise in space-based ISR, communications, and METOC essential to fielding the capabilities required by Sea Power 21.

In general, there are two categories of space S&T activity in need of Navy support. The first includes the basic and broad S&T development underlying most space programs. The Navy needs to maintain a vigorous basic research program so that it will continue to ensure its ability to participate in, and benefit from, the full range of space mission areas and activities. This program should support not only the technology per se, but also the analytic skills needed by the naval space cadre to enable it to act effectively in joint program management. Much of this in-house space technology expertise now resides at NRL's Naval Center for Space Technology (NCST).

The second category of space-related S&T in need of Navy support includes specific technologies important to fulfilling this Service's unique space needs. As noted earlier, such specialized needs include the space-related technologies associated with ocean surveillance, oceanic and littoral environmental conditions, mobile communications, and navigation.

One significant endeavor in the Navy's support for space S&T is the launch of the Coriolis spacecraft with the WindSat sensor onboard. The WindSat sensor, a joint Navy and NOAA project, is likely to contribute significantly to wind vector measurement over the world's oceans and to enable better ocean wind vector measurements from future NPOESS satellites. Continued Navy support for innovative and well-integrated space experiments such as this is seen as a positive role for future Navy activities. On the negative side, the Office of Naval Research (ONR) did not support WindSat, and in fact ONR recently abolished the Remote Sensing and Space Office within its Ocean, Atmosphere and Space Department. The closing down of ONR's remote sensing office concerns the committee, as it seems to indicate an intent not to support this research area. Research efforts in the area of remote sensing and space are clearly needed to reinvigorate the Navy's space S&T activities.

The Navy's base of space technology expertise is substantially housed at NCST.[12] NCST has historically received little support from ONR or other naval offices (beyond NCST's core NRL support); rather, most of NCST's program support has been provided by non-naval organizations such as the Missile De-

[12]See Appendix A for a detailed history of the Navy's involvement in space.

fense Agency (MDA), NOAA, NRO, and others. This funding has allowed NCST to maintain a critical mass of personnel, facilities, and technical credibility and has resulted in the development of a number of capabilities useful to the Navy (such as WindSat). In addition, NCST recently initiated or proposed several other programs focused on experimentation with direct user tasking, naval-platform-compatible wideband communications, improved radio-frequency (RF) emitter tracking, and hyperspectral imaging capabilities that may have the potential to significantly enhance naval operations.

One such project is NCST's current Tactical Microsatellite (TacSat) effort, aimed at testing concepts of operation for providing tactical field commanders direct satellite control. In particular, the first TacSat will test the ability for an airborne Navy EP-3E aircraft to control an on-orbit RF collector, to receive the RF data in real time, and to synthesize the satellite and EP-3E RF data to demonstrate an improved RF emitter geolocation capability. Future TacSat plans are to build on this test to investigate direct user tasking of electro-optic/infrared (EO/IR) and other space-based capabilities using a variety of tactical platforms. Thus, NCST's TacSat effort, which currently receives little Navy funding,[13] could prove the utility of a wide range of S&T and operational capabilities that are of current interest to the Navy. The Navy may be well served by increased interaction with NCST and financial support to expand NCST's development of maritime space mission concepts, on-orbit demonstrations of space-based capabilities that support maritime operations, and transition of technology to industrial or government partners for system production.

An overall concern exists that the Navy's support for space S&T has atrophied in recent years, thereby jeopardizing the Navy's ability to meet its future responsibilities. Indications of this problem include the lack of ONR program support to NCST referred to above, the lack of Navy support for innovative space mission concepts such as TacSat, the recent cancellation of the NEMO and GIFTS satellites, and the successful launch of Coriolis-WindSat. While Coriolis is a successful program, its launch has reduced the Navy's current support of satellite systems development.

Each of these four satellite programs has been supported at a level of (or was projected to need) $10 million to $20 million annually through satellite construction and launch. While the three Navy-funded satellites (Coriolis, NEMO, and GIFTS) received strong support from the Navy as well as from partner agencies, each system experienced a different fate: Coriolis was launched successfully in 2003; NEMO was cancelled in 2002, apparently following financial troubles with its industrial partner; and GIFTS was cancelled in 2003, apparently following a

---

[13]Current Navy funding for TacSat is limited to support from NRL for initial program development, Navy-TENCAP for the signals payload (an airborne experimental payload space-qualified by NRL), and the Naval Air Systems Command for the needed modifications to four EP-3E aircraft to be used during the TacSat evaluation and testing phase.

funding-related scheduling setback and the loss of its Space Test Program (STP) launch date. As a result of these recent events, the overall Navy METOC space budget has fallen from $22 million in FY03 to a request for $4.2 million in FY05,[14] significantly impacting the Navy's ability to support future Navy METOC needs.[15] This lack of support for space-related S&T can be attributed in part to the lack of a clear focal point for space-related program and budget matters on the staff of the Chief of Naval Operations and to a lack of the cost-benefit analysis necessary to prioritize and protect sufficient funding for these programs.

The problems with Navy support for science and technology have also been compounded by an artificial distinction within the Navy S&T budgeting process: the distinction made between S&T programs that directly support major planned future Navy acquisition programs (of which there are few for space) and those that do not—with the latter receiving minimal funding. This narrowly construed Future Naval Capabilities (FNC) process has inadvertently helped to divest direct support from naval space technology. This weakening of financial support has also resulted in the diffusion of effort at NCST into activities that are not in the mainstream of the Navy's responsibilities (for example, currently NCST receives a majority of its financial support from non-Navy sources). The result has, in effect, perversely required Navy space S&T ideas to undergo a more difficult granting process than, for example, that for ship and aircraft S&T. In that area, in contrast to Navy space programs, well-funded ongoing acquisition programs enable transition-related S&T funds and also provide a direct source of technology development funding. This situation may have led, in FY03, to ONR's supporting $10.7 million in basic research funds (6.1) for the study of space environmental effects and $3.3 million in applied research funds (6.2) for the study of spacecraft technology, but supporting no advanced technology development (6.3) projects in these areas.[16] These figures are in contrast to the Navy's overall 6.1, 6.2, and 6.3 budget of $1.6 billion in FY03.[17]

Recent METOC program cancellations (NEMO and GIFTS, which had been supported through 6.3 funds) appear to indicate a lack of continuing support for the development of novel spacecraft meeting maritime needs. The committee believes that relatively small additional Navy investments in space S&T (as compared with the Navy's entire S&T funding) provided directly or through other organizations, such as DARPA, could provide the technology and expertise needed to ensure that future naval operations are effectively supported by new, non-Navy space systems and programs.

---

[14]Funding data derived from recent DOD budget appropriations for FY98 through FY04.

[15]See the section entitled "Meteorology and Oceanography" in Chapter 4 for further details.

[16]RADM Jay Cohen, USN, Chief of Naval Research, presentation to the committee, July 29, 2003.

[17]Office of the Under Secretary of Defense (Comptroller). 2002. "RDT&E Programs (R1)," *Department of Defense Budget, Fiscal Year 2003*, Department of Defense, Washington, D.C., February, p. N-1.

Recommendation 3.3. The Chief of Naval Research (CNR) should maintain a critical level of space mission area funding aimed at supporting current maritime needs as well as at providing broad support to base-level technologies with the potential to support National Security Space programs, such as the Transformational Communications Architecture and Space Based Radar programs.

Specifically, the CNR should continue or increase current levels of basic research (6.1) and applied research (6.2) funds in support of space technologies and systems. In addition, the CNR should consistently allocate advanced research and development (6.3) funds to enable regular Navy space sensor development and on-orbit testing. Given recent Navy space sensor program allocations as a benchmark, the committee envisions that a level on the order of $40 million annually of 6.3 support would be sufficient to ensure regular development of new Navy space systems. Specific space mission areas recommended are as follows:

- *Communications.* Robust on-orbit capabilities supporting naval communications needs. These range from connecting Global Information Grid-enabled future sea-basing command centers to critical low-latency weapons control connections necessary for effective missile defense;
- *Intelligence, surveillance, and reconnaissance; and meteorology and oceanography.* Improved on-orbit capabilities, ranging from modified sensors to new capabilities including hyperspectral imaging, to support emerging needs arising from Sea Power 21; and
- *Data fusion.* Using space-derived information and systems in scaling and optimizing global information capabilities in support of Sea Power 21 operations.

Experimentation

The transformation of space systems by the DOD and the intelligence community will continue to offer new opportunities to improve naval capabilities. One ready means for taking advantage of available opportunities is for the Navy to engage in forward-looking experimentation efforts. Typically, experimentation is used not only to test new technologies that may prove useful to the operational forces, but also to provide a forum in which operators can practice with developing systems and thus study how systems could be modified so that they could be leveraged into programs of record. Additionally, experimentation supports the Navy's overall operational analysis community and often identifies, or highlights, capability gaps in need of further developmental S&T work. A key to the experimentation process, however, is a clear identification, up front, of operational needs. Only then can experimentation efforts be constructed to appropriately address the Navy's needs.

Current Navy experimentation programs, however, do not appear to be derived from articulated operational needs and thus do not appear to effectively support the development of new capabilities. In part this lack appears to have resulted from the Navy's not supporting rigorous operations analysis of the missions and capabilities necessary to accomplish the goals for Sea Power 21. The Navy has acknowledged the need to tie operational needs more effectively into experimentation efforts, which has resulted in the establishment of Sea Trial. As discussed in the Sea Trial Campaign Plan, Sea Trial establishes the "process to go from strategy based concepts through experimentation to proposed [mission capability plans] and the [naval capability plan], to changes in doctrine, organization, training, material, leadership, development, personnel, and facilities (DOTMLPF)."[18] While led by the Commander, Fleet Forces Command, the Sea Trial effort is managed by the Navy Warfare Development Command (NWDC), and for issues related to space and network-centric operations it is also guided by the Naval Network Warfare Command.[19] The Sea Trial plan, while involving space systems, does not include support to produce and launch new experimental space capabilities; rather, it focuses on the exploitation of existing or planned systems or on the use of airborne surrogates for potential space technologies.

Another new driver in terms of the use of experimentation is the ever more rapid advancement of technology. This speed of progress makes it increasingly important to allow warfighters to experiment with new technology and to explore how to use that new technology to accomplish operational missions more effectively or efficiently. To support rapid development and experimentation, the DOD, in 1995, created the Advanced Concept Technology Demonstration (ACTD) program. The ACTD program has been successful in allowing military operators to experience new technology, and it has also allowed researchers and operators critical feedback on how best to use the technology in real-world operational settings. One such recent ACTD program was Global Hawk, a high-altitude, unmanned surveillance aircraft that is now receiving strong interest from the Navy. The ACTD program, used as a technology supplier for experimentation, thus also allows technology to advance through a spiral development process, often involving the use and modification of commercial-off-the-shelf components, toward the adoption of new technology into the Services.

Past Navy efforts in space experimentation are well documented, going back to the 1950s when Navy efforts achieved significant breakthroughs in satellite development (Vanguard), characterization of ocean features (the Geodetic Satellite—Geosat), navigation (Transit and Timation), and ocean surveillance.[20] Much of this experimentation, supported through NRL as well as other organizations,

---

[18]Commander, U.S. Fleet Forces Command. 2003. *Sea Trial: Concept Development and Experimentation Campaign Plan (U)*, Norfolk, Va., June 30, p. 11 (an unclassified excerpt from a classified document).

[19]The Naval Network and Space Operations Command provides operational input to NETWARCOM on space-related experimentation issues.

[20]See the more detailed history of the Navy's development of space provided in Appendix A.

was not necessarily focused on unique naval space needs and has had application to many military and civilian elements. For example, the Navy's historical capabilities in space-based navigation and timing are still recognized today in the maintenance of the DOD timing standards. In addition, NRL performs the on-orbit clock analysis for all of the GPS satellites, houses the GPS Joint Test Agency working group leading time and frequency characterization for all GPS receivers, supports new GPS clock development, and (in concert with the Space and Naval Warfare Systems Command (SPAWAR)) had the lead role in developing the only military-approved satellite GPS receiver.

The Navy has a strong history in space experimentation involving ocean surveillance from space, both in open-ocean and littoral regions (choke points) to monitor movements of all types of ships, and it has led the way in the application of national assets in the detection of high-performance aircraft that could threaten deployed naval forces. These efforts have also extended into the development of METOC satellites, including WindSat, which monitors sea-surface wind speed and direction, and Geosat, which provides precise measurements on sea-surface height.

In recent years the Navy's experimentation programs involving space have been largely opportunistic, taking advantage of available new technologies to make incremental improvements in fleet performance. These programs have focused heavily on antenna technologies for mobile platforms. Navy-TENCAP is a commendable example of experimentation with fielded NSS programs (ISR in particular), enabling those systems to provide critical intelligence to the tactical users. Today, a benchmark for tactical exploitation is the Navy's use of the national ELINT systems. Unfortunately, current Navy space system experimentation is weak and does not appear to have come from a strategic and tactical planning process that identified and prioritized new capabilities necessary to meet the goals of Sea Power 21 successfully. As a result, there appears to have been limited success in transitioning successful space experimentation results into programs of record.

Recommendation 3.4. As part of the Sea Trial experimentation process, the Commander, Fleet Forces Command, should formalize the roles between the Naval Network Warfare Command and the Navy Warfare Development Command pertaining to maritime and joint forces experimentation in space and space-related areas so as to fully exploit and complement the Joint Forces Command experimentation process and to explore the best uses of future space-based intelligence capabilities.

In particular, NETWARCOM and NWDC should carry out the following:

- Coordinate with the Deputy Chief of Naval Operations for Warfare Requirements and Programs in order to generate experimentation initiatives aimed at addressing space capabilities requirements;

- Perform analysis, modeling, and simulation in a simulation-based acquisition approach on potential new space capabilities before proceeding to testbeds and field experiments; and
- Conduct experimentation aimed at supporting new or improved sensors and subsystems that can piggyback on available NSS satellites.

Personnel/Space Cadre

An expanded naval space cadre is the key to ensuring that naval equities and needs can be articulated, addressed, and satisfied under the new National Security Space/DOD Executive Agent for Space structure. To this end, the Department of the Navy will need to develop and maintain a sufficient cadre of space-qualified personnel to support its component of national space planning, programming, acquisition, and operations.

Navy efforts in the identification and management of the Navy space cadre started well before the formal establishment of the DOD Executive Agent for Space, and significant initial progress has been made. Although the Navy space cadre will represent only a small percentage of the total number of DOD space personnel (who are primarily Air Force personnel), experienced and motivated naval personnel, aggressive management, and targeted assignments across all space mission areas can provide considerable leverage in the effort to satisfy Navy needs in space. The recently adopted procedure of highlighting the need for space expertise to selection boards should be formalized and continued, to ensure that these uniquely experienced personnel are promoted and retained. The scope of involvement envisioned for the space cadre, combined with the potential impacts on naval systems and operational concepts, argues strongly for a significant expansion of space-coded billets and qualified personnel.

The Navy's current strategy for the naval space cadre envisions (1) a diverse blend of officers, civilian, and enlisted personnel rather than a separate community; and (2) appropriately coded billets in all space-related functional areas. In the Navy, there are approximately 250 space-coded officer billets, with approximately 700 officers identified as possessing the basic subspecialty designation. Little progress has been made in identifying either space-related billets or personnel for the enlisted and civilian components of the cadre. In addition, by the end of FY03, only approximately 20 percent of the space-coded billets had been filled by members of the Navy space cadre.[21] It is anticipated that in time, as the space cadre is strengthened, this percentage will increase.

A credible and sustainable space cadre starts with advanced education. The Naval Postgraduate School (NPS), through its space systems curriculum, is a recognized center of excellence for providing graduate-level education to officers

[21]RADM Rand Fisher, USN, Commander, SPAWAR Space Field Activity, and Director, Naval Space Technology Programs, discussions with the committee, October 29, 2003.

of all of the military Services, as well as to selected DOD civilians. Relative to the trends of the previous decades, a disturbing decrease was noted in the recent Navy quotas and assignments to the space systems curricula. Given the opportunities, challenges and responsibilities represented by the new DOD Executive Agent for Space structure, this adverse trend concerns the committee. In contrast, it was noted that the entry of Marine Corps officers into the NPS program as well as Corps contributions in the design and content of the NPS curricula are increasing. These developments represent the Marine Corps establishment of a space cadre and its growing efforts to influence NSS programs that can meet Marine Corps needs.

The process of implementing the new directive establishing the DOD Executive Agent for Space structure has identified both opportunities and issues relative to the future direction of the NPS space curricula. Closer interaction and coordination with the Air Force Institute of Technology is anticipated in such areas as curricula scope and duration. Naval leadership must remain fully engaged in the DOD Executive Agent for Space-led curricular oversight process to ensure that NPS curricula content and quality remain fully responsive to naval educational skill requirements.

Recommendation 3.5. The Chief of Naval Operations should strengthen and expand the Navy space cadre as follows:

- Continue formalizing the leadership of the Navy space cadre under the Deputy Chief of Naval Operations for Warfare Requirements and Programs;
- Provide additional (new) billets to support National Security Space (NSS) research, development, and acquisition efforts;
- Ensure opportunities for positions of responsibility in all NSS activities and space mission areas;
- Review the function of fleet and operational staff billets and assign space codes to billets as appropriate; and
- Reexamine the Navy's support of and quotas for the Naval Postgraduate School space systems programs in light of expanded naval involvement in NSS activities.

4

Implementation: Navy Support to Space Mission Areas

Naval forces have continued to be major users of information derived from space-based systems of all types. In order to help identify the Navy's needs in space for providing future capabilities, the committee reviewed the Navy's participation and needs with respect to current, planned, and proposed space-based capabilities that may have an impact on the successful implementation of Sea Power 21 warfighting concepts. Specifically, the committee's review is structured around these space-based capabilities and the six National Security Space (NSS) mission areas: intelligence, surveillance, and reconnaissance (ISR); meteorology and oceanography (METOC); theater and ballistic missile defense (TBMD); communications; position, navigation, and timing (PNT); and space control.

Table 4.1 provides a status of several current, planned, and proposed space-based capabilities, from which one can conclude that the Navy needs to collaborate with a significant variety of agencies outside the Navy in order to ensure that its Navy-unique needs are satisfied. Thus, the Navy's participation in the development of space systems needs to be flexible in order to meet the needs of a variety of partner agencies as well as to enable effective leadership of the Navy's own programs and initiatives. The sections below detail the Navy's current participation across each of the space mission areas and also provide an assessment of how the Navy can improve its use of these systems and better influence the development of new systems, thus ensuring that its needs in space will be met in the future.

INTELLIGENCE, SURVEILLANCE, AND RECONNAISSANCE

The Navy has recently been moving toward the adoption of Sea Power 21 as a global instead of a regional strategy, but a strategy that still addresses regional and transnational threats. These threats are unlikely to be concentrated in a few regions that can be simultaneously addressed using concentrated forces. Instead, the National Security Strategy states that it will be necessary to address threats anywhere on the globe, and at a tempo that will permit dealing with many, widely dispersed threats quickly, decisively, and nearly simultaneously.[1] This calls for forces to operate using the most exact information possible about the enemy, to analyze that information to determine critical nodes to attack, and to direct weapons systems, launched from widely dispersed platforms, to strike those nodes. The role of the Navy in such a strategy is especially important, given its traditional forward presence, sea dominance, and strategic sealift. In this role, the Navy capitalizes on and builds from NSS and organic Navy ISR capabilities to support broad coverage and over-the-horizon (OTH) targeting, followed by precision strike using precision-guided munitions (PGMs).

The capabilities of the individual Sea Power 21 pillars—Sea Strike, Sea Shield, and Sea Basing—are all, to varying degrees, dependent on ISR involving data from space-based, airborne, ground-based, and sea-based sensors. These capabilities are also central to the realization of the FORCEnet foundation that enables the operational implementation of Sea Power 21. NSS systems have proven invaluable for both complementing and expanding the ISR capabilities provided from other sources. ISR information from NSS sensors is merged with ground-, sea-, and air-based sensor data to develop fused products to assist in the positioning and repositioning of platforms and forces; to identify and assault critical vulnerabilities and centers of gravity of enemy forces; to assess damage; and to permit the efficient use of resources to conduct concurrent and follow-on missions. Furthermore, NSS support is used to deter enemies from employing an effective threat to U.S. strike operations by maintaining surveillance capability against potential conventional as well as unconventional strikes.

NSS ISR capabilities could become significantly more important to the Navy if, in the future, near-real-time persistence from NSS ISR systems was achieved. For space systems, persistence is achieved by increasing either the period of observation by sensors on a particular satellite (for instance, geosynchronous satellites can continuously observe one-half of Earth at a time) or by increasing the number of satellites carrying the particular sensors (for instance the Global Positioning System (GPS) uses a constellation of 24 satellites to enable continuous ground observation of at least 4 GPS satellites).

---

[1]President George W. Bush. 2002. *The National Security Strategy of the United States of America*, The White House, Washington, D.C., September 17.

TABLE 4.1 Status of Current, Planned, and Proposed Department of Defense Space Systems Programs Including General System Limitations and Risk

Space-Based Capability	Current Programs	Available?
Imagery (infrared, visible, radar)	NRO systems, FIA, SBR commercial imagery	Yes
Electronic intelligence (ELINT)	NRO systems	Yes
Navigation	GPS	Yes
Timing	GPS	Yes
Meteorology and oceanography	GOES, POES, NPOESS	Yes
Ground moving target indication	SBR	No
Airborne moving target indication	None	No
Boost-phase missile defense	SBIRS-H	No
Midcourse missile defense	SBIRS-L	No
Space-based IP networks (GIG)	TCA	No
Satellite communications	MUOS, MILSTAR, AEHF, commercial	Yes

NOTE: A list of acronyms is provided in Appendix G.

The primary constraint on ISR satellites is related to simple physics: orbits enabling long duration over a spot on Earth require high-altitude orbits—but the higher the orbit, the lower the fundamental resolution of any orbital imaging system. Thus, high-resolution space imaging systems require a large number of less-expensive, low-Earth-orbit satellites, or a smaller number of medium-Earth-orbit or geostationary-Earth-orbit satellites carrying extremely expensive high-resolution imaging systems. Recently, several novel schemes have been proposed for using fleets of microsatellites linked as an interferometric array, thus providing high resolution with small identical satellites. Such proposals are in the very

Oversight Agency	General Limitations	Status/Risk
DOD Executive Agent for Space, NRO, NGA	Satellite revisit time	FIA and SBR in development, SAR versus GMTI trade-offs for SBR
DOD Executive Agent for Space/NRO	Encryption, geolocation accuracy	Gap-filler satellite may fill need
Air Force	Vulnerability to countermeasures	Enhanced jamming protection programmed
Air Force/Navy	Few	Ongoing clock development
NOAA	Passive sensors only, resolution and revisit times, international partnerships	Cost, feasibility of active sensors
Air Force	Revisit time, field of view, data rates	R&D issues, cost, trade-offs between SAR and GMTI, ubiquitous coverage
None designated	Stressing technology, data rates	Ubiquitous coverage, use of space sensor for weapon guidance
MDA	Stressing technology	Cost, technical risk, system under development
MDA	Stressing technology	Cost, technical risk, no current program
DOD Executive Agent for Space	Wideband laser links	Development risk, cost, system under development
Air Force and Navy	Data assurance, link availability	Bandwidth demands growing rapidly

early development stage and might represent a novel route for additional science and technology (S&T) support. Combining improved persistent NSS ISR capability with improved communications, processing, and exploitation systems will enhance the ability of the Navy to engage in future missions (provided that the capabilities are developed in accordance with naval needs).

One additional factor (discussed in the major section below entitled "Space-Based Communications") is the need not only for timely acquisition of ISR information, but also for its timely analysis and dissemination. This need has recently been embodied by the Department of Defense (DOD) in the information dissemination concept known as TPPU, or "task, post, process, and use." Under TPPU, all ISR data are to be made immediately available (posted) for processing and use by any potential end user. This concept has the goal of making ISR data available for use more quickly than has been possible in the past. However, under TPPU large volumes of unprocessed (raw) ISR data will be regularly transmitted to end users, and the committee is concerned that TPPU may significantly increase the communications bandwidth needed to effectively access and utilize the ISR data.

Future Space-Based Intelligence, Surveillance, and Reconnaissance Systems

Historically, NSS ISR systems have been managed by the National Reconnaissance Office (NRO) and have significantly improved the effectiveness of naval missions. The Navy has been a major participant in these NRO programs to ensure that naval interests are served. While these past and current NSS ISR systems and augmentations have proven their value, NSS systems currently in the development and/or planning stage by the NRO and the Air Force hold promise of even more improvements in naval capability. Two systems in particular are noteworthy in this regard:

- The Future Imagery Architecture (FIA) being developed by the NRO, and
- The proposed Space Based Radar (SBR) being planned by the Air Force.

FIA is being planned to replace the NRO's existing series of national optical and infrared imaging satellites. While the details of the program are largely classified, the FIA initiative will represent a significant improvement to the nation's space-based imagery systems, in terms of both resolution and persistence. The Navy needs to remain engaged in this initiative to ensure that its maritime imagery needs will be met by FIA.

SBR represents a plan to field a space-based radar to enable near-continuous monitoring (through either radar imagery or ground moving target indication (GMTI)) of the majority of Earth's surface—a critical supporting capability to enable the Navy to provide maritime domain awareness consistent with its role in homeland defense. SBR is such a major ISR initiative that the United States is unlikely to support an alternative space-based radar initiative in the next several decades. Thus, it is imperative that the Navy understand its needs for this extended time period in order to be able to ensure that its needs will be met by SBR. While the capability assessment for NSS ISR (summarized in the tables in Appendix C) is based on what could be provided with a nominal SBR, the committee believes that the Navy has not yet interacted with the SBR program office on the scale required to assure that Navy requirements will ultimately be met. The few individuals currently representing the Navy's interests in SBR have made significant progress toward inserting Navy requirements into the SBR planning process. These inputs, however, appear to have been based on limited background analyses.

The Navy is developing an understanding of the operational implications of the various modes and capabilities of SBR against future naval scenarios. The Air Force has agreed to conduct some simulations involving near-term maritime scenarios, but SBR also needs to be integrated into maritime modeling and simulations looking over the several decades that SBR would be in existence. As discussed elsewhere in this report,[2] the Navy currently lacks a modeling and simulation effort to support thorough operations analysis that can permit cost-benefit and performance trade-offs across a wide range of systems (including SBR) against the needs arising from a wide range of future naval scenarios. Without full and traceable analysis of the requirements for supporting Sea Power 21, including a good understanding of the related technical issues, the committee is concerned that SBR may not meet some of the Navy's key performance and operational requirements.

Because there has been little Navy S&T funding for SBR, there has, to this point, been an incomplete understanding of numerous aspects of how a maritime SBR could be designed differently from the current SBR baseline. To overcome some of these deficiencies, the Navy recently funded the Naval Research Laboratory (NRL) to identify new SBR modes of operation that can support maritime operations. As of early December 2003, the Air Force had accepted some of these options into the SBR baseline. In this case, the Navy was able to leverage its internal S&T experience to help define the technical features of the SBR needed by the naval forces. The committee encourages further similar Navy involvement.

The FIA and SBR programs will provide improved capability for monitoring fixed targets and threats (both FIA and SBR) and moving targets and threats (SBR). Following is an assessment of how existing and planned NSS ISR systems enable the capability areas identified by Sea Power 21. The resulting Sea Power 21 capability dependencies are summarized in Appendix C.

Sea Power 21 Capabilities

Sea Strike

Through Sea Strike operations, naval forces will execute and direct decisive and sustained influence in joint campaigns. Sea Strike relies on a combination of ISR assets—space-based, theater, and force—to support the conduct of the following types of operations:

---

[2]See the section entitled "Navy Space Support" in Chapter 3.

- *Strike operations* against fixed and moving targets on land and on sea,
- *Special operations* that require precision targeting information, and
- *Defensive operations* necessary to ensure strike aircraft survivability.

Today, strike targets are identified, classified, tracked, and geolocated through a combination of sensors on NSS systems, airborne platforms, and naval platforms. NSS and airborne systems are generally used cooperatively to support time-sensitive requirements of strikes. The requirements of the Navy for overland targeting are essentially identical to those of the other Services; however, the Navy will need to carefully manage and guide the course of progress on its requirements for over-water targeting to ensure that they are included in future programs. In particular, many satellite systems do not operate over the open ocean (this includes early plans for the SBR described above)—pointing out the Navy's need to track even its most basic requirements on availability. During a system's development and operational phases, technical and funding support is typically needed to improve performance and adapt the system to changing threat and target conditions. Additionally, the Navy will need to explore the potential of other new space ISR capabilities, such as hyperspectral imaging to assist in separating targets from background and camouflage, especially in the open-ocean and littoral areas unique to the operations of Navy and Marine Corps forces. In general, the future FIA and SBR systems could greatly enhance NSS support for Sea Strike by—

- Improving persistence through increased numbers of satellites, and
- Improving image resolution, thereby strengthening the ability of naval forces to identify, track, and target terrorist and other small-unit threats.

Sea Shield

To maintain littoral superiority for naval and joint force components, ISR resources must be able to support protection against conventional and unconventional (i.e., chemical, biological, radiological, nuclear, and environmental) threats from special operations and terrorist forces. Information from space-, ground-, and sea-based and airborne ISR resources need to be used, where possible, to identify and locate near-horizon and over-the-horizon threats, to enable afloat operations by supporting self-defense against and/or neutralization of undersea threats (including those from submarines, mines, submerged barriers, and obstacles), and to provide defense over land and over sea against theater air and ballistic missile threats. The support of all of these defensive operations currently challenges NSS ISR resources and will continue to do so for the foreseeable future.

One of the limitations of current NSS systems in contributing significantly to defensive antisubmarine warfare (ASW) and countermine operations is the lack of persistence in making observations of offensive enemy operations. It is possible to observe enemy submarines at shallow depth from space, and also to observe the laying of mine fields or the navigation by enemy combatants through mine fields they have laid. However, the long time lapses between overhead satellite observations by current NSS systems do not support the near-continuous observations needed.[3] As described above, the future FIA and SBR systems, if fielded, should significantly improve overall observational persistence.

Today, most operations rely largely on theater assets (the SPY-1D radar system on the Navy's Aegis ships, sensors on E-2C and E-3 aircraft, and so on) to provide the ISR information necessary to support Sea Shield operations effectively. For surface warfare, Sea Shield requires that ISR capability provide near-horizon and over-the-horizon warning, tracking, and targeting information against surface targets; these requirements are similar in many regards to the Sea Strike capability needs. In addition to the improvements noted above that would enhance NSS support for Sea Strike, the future FIA and SBR systems should greatly improve NSS support for Sea Shield by—

- Increasing coverage areas, thereby extending the engagement distance to distances beyond the threat range from enemy combatants; and
- Establishing a space-based GMTI capability (with SBR), thereby enabling space-based, near-continuous tracking of moving surface vessels.

Similarly, undersea warfare support can be extended in area by improved persistence of SBR and FIA, provided that these systems are designed and operated specifically to address the special needs of large-area search in ocean areas. These forms of support are just the beginning, however, and long-term S&T is needed in support of effective naval specification and use of SBR. As an example, further S&T funding could be provided to support a comparison of the expected performance of radars with which the Navy is familiar (such as the E-2C aircraft radar and its upgrades) with the various options for SBR. Such analysis would help establish and maintain the connection between specialized maritime radar experts, the operational Navy, and the SBR office.

Sea Basing

Sea Basing entails the provision of the full capabilities of an at-sea joint command center as well as the provision of all of the associated sea-based logis-

[3]The Office of Naval Research has developed, under its littoral remote sensing effort, a series of automated analysis algorithms for using data from the nation's space-based intelligence assets to assist with nearshore mine and mine field detection from the surf zone on to the beach. These algorithms have transitioned, via the organic mine countermeasures Future Naval Capabilities program, to the Naval Oceanographic Office Warfighting Center. Fleet awareness and use of these algorithms are continuing issues.

tics needed for initial surface and amphibious strike actions. Thus, the ISR needs for Sea Basing are covered primarily by providing access to the capabilities needed to support Sea Strike and Sea Shield. The additional capabilities needed for Sea Basing are primarily structured around logistics support needs (primarily for optimal ship route planning) and are described in the next major section, "Meteorology and Oceanography."

FORCEnet

Force projection and defense from forward-deployed Navy platforms depend on the efficient networking of naval, national, and joint nodes involved in all aspects of information production, command responsibility, and control authority through communications and computing power to meet Navy objectives. These topics are broadly included under the heading of FORCEnet.

ISR derived from NSS sensors is an important portion of FORCEnet, and the information from these NSS sources will be used in conjunction with information gathered from other sources to support overall Navy objectives. In reality, most strike actions and most projections of defensive capability will cooperatively use information from multiple ISR sources to engage targets and threats, and the actual source of any individual information element will be transparent to the warrior. The timeliness, relevance, quality, and quantity of the information, however, will continue to determine the outcomes of naval missions. As the mission of the Navy grows in the 21st century and as the theater of importance expands to global dimensions, the importance of access to global, persistent ISR information will necessarily grow. In other words, as the need for technology associated with space grows, the technology associated with the analysis, exploitation, and fusion of NSS-derived data will also need continued improvement.

In summary, then, it is unlikely that the Navy will be able to meet its overall Sea Power 21 ISR needs without a strong program of S&T, space engineering, and program participation involving naval and Navy-sponsored personnel. These efforts will also require a continuing commitment of personnel and funding to NSS ISR activities.

Capability Shortfalls and Technology Gaps of National Security Space Intelligence, Surveillance, and Reconnaissance Systems for Navy Use

Given the needs discussed above for ISR support from space, there are several shortfalls and gaps in current and currently planned NSS ISR capability that will limit the Navy in carrying out the elements of Sea Power 21 effectively. These are summarized in Table 4.2.

The approaches that the Navy has available to it to address the shortfalls and gaps listed in Table 4.2 fall into four general areas:

TABLE 4.2 Shortcomings of National Security Space (NSS) Intelligence, Surveillance, and Reconnaissance (ISR) Systems for Navy Use

NSS ISR Shortcoming or Gap for Navy Use	Navy Issues
Persistence	Long revisit gaps between images, no long dwell sensors, too few satellites, and no moving target indication (MTI) radars. Space Based Radar (SBR) and Future Imagery Architecture (FIA) may alleviate gap.
Area coverage	Operation of satellites over ocean, resolution versus coverage, synthetic aperture radar versus MTI look angles. SBR and FIA may alleviate gap.
Ocean operations	Operation over oceans, operating modes to counter effects of ocean clutter. SBR and FIA may alleviate gap.

- *Supporting S&T to address the NSS ISR shortfalls and gaps and then transitioning the results of this S&T work into planned and ongoing NSS programs.* This approach requires a knowledgeable space cadre that understands both space technology and the operations of naval forces. To date, the Navy has used such an approach effectively in many programs for which the NRO or the Navy itself has been the lead agency. Key to this success has been the Navy's Technical Exploitation of National Capabilities Program (Navy-TENCAP), which is specifically designed to provide a link between existing national capabilities, naval operations, and fleet experiments. The Navy can benefit from a similar program aimed to experiment with Air Force programs.
- *Directly participating in programs that can mitigate these NSS ISR shortfalls during the course of the programs' evolution.* The involvement of the space cadre in these programs needs to be carefully coordinated, and billets might sometimes need to be accompanied by modest Navy dollars to provide the leverage necessary to support Navy direction and to promote changes sought by the Navy. Again, the Navy has used this type of approach well in programs with the NRO, and it would be well served to expand this level of participation into most Air Force programs.
- *Providing significant Navy dollars to relevant space programs in order to augment funds provided by the intelligence community and the DOD.* These Navy funds would support changes to planned or developing programs to ensure that capabilities needed primarily by Navy users are included in NSS IRS systems.
- *Supporting the development and testing of other novel sensors, such as the hyperspectral Naval EarthMap Observer (NEMO) satellite.* Such hyperspectral systems have shown great potential to support both METOC and ISR needs.

The creation of NSS ISR systems to provide capabilities needed by the Navy would clearly be valuable and perhaps even essential to the full implementation of Sea Power 21. Such systems can be envisioned, but many represent a technical challenge, given today's state of technology.

Recommendations Regarding Intelligence, Surveillance, and Reconnaissance

Recommendation 4.1. The Department of the Navy should develop and fund directed operational analysis and science and technology (S&T) programs focused on addressing the Navy's intelligence, surveillance, and reconnaissance shortfalls independent of whether or not the affected programs are managed by the Department of the Navy or the Department of Defense (DOD) Executive Agent for Space. The Department of the Navy should also work to transition the results of these efforts into planned and ongoing National Security Space programs.

Recommendation 4.2. The Department of the Navy should continue its full support of National Reconnaissance Office intelligence, surveillance, and reconnaissance (ISR) activities and seek to extend its involvement in ISR program planning, development, and execution across other agencies' ISR efforts.

Recommendation 4.3. The Department of the Navy should provide budget authority to augment National Security Space intelligence, surveillance, and reconnaissance programs to permit program and system additions that address needs unique to Navy strategies (such as maritime operation).

Recommendation 4.4. The Department of the Navy should coordinate with other agencies to support the development of advanced sensing technologies not currently part of the program plans of the DOD Executive Agent for Space. One such program that the committee believes has significant potential to provide new naval capabilities is the Naval EarthMap Observer (NEMO) hyperspectral imaging satellite.

METEOROLOGY AND OCEANOGRAPHY

The Navy has funded remote sensing research and development (R&D) in the areas of meteorology and oceanography nearly since the beginning of its involvement in space, but the military's role in developing METOC satellites is changing. A 1994 Presidential Decision Directive mandated that the national military and civilian METOC communities consolidate all METOC satellites and systems under the direction of the National Oceanic and Atmospheric Administration (NOAA).[4]

Many of the Navy's current METOC efforts are coordinated by the Oceanographer of the Navy and the Commander, Naval Meteorology and Oceanography Command (NMOC). The two largest centers within NMOC are the Naval Oceanographic Office (NAVO) and the Fleet Numerical Meteorology and Oceanography Center (FNMOC). Each of these offices also has a collocated NRL R&D establishment to support its mission. NAVO, the largest single element of the Navy's METOC commands, is one of the Navy's two primary METOC analysis centers. FNMOC is the DOD's principal operational processing center for automated numerical METOC analyses and predictions; as such it provides a continuously updated METOC picture for use by the DOD.

NAVO is headquartered at the National Aeronautics and Space Administration's (NASA's) John C. Stennis Space Center near Bay St. Louis, Mississippi. Its primary mission is to conduct oceanographic multidisciplinary surveys in the world's oceans. The office collects hydrographic, magnetic, geodetic, chemical, navigation, and acoustic data using ships, aircraft, spacecraft, and other platforms. In addition, NAVO provides much of the ground-based data necessary to calibrate and monitor the performance of the Navy's remote sensing systems.

In the past, the Navy has funded several satellite systems for Navy-unique METOC applications.[5] These include the Geodetic Satellite (Geosat) and its successor Geosat Follow-on (GFO), NEMO, Geosynchronous Imaging Fourier Transform Spectrometer (GIFTS)/Indian Ocean METOC Imager (IOMI), and Coriolis-WindSat (see Box 4.1). Geosat and its successor GFO are radar altimeter satellites, developed primarily to map sea-surface heights. Such data are used to augment ocean circulation models and to help predict ocean weather. In addition, GFO data are used to refine the DOD geophysical models used by the ballistic missile submarine community. To date, the Navy has invested approximately $100 million in the GFO program and is planning to begin architectural studies related to development of a next-generation altimeter system.[6] Such efforts are strongly encouraged since, following the design lifetime of GFO (its 5 year mission will end in 2005), the Navy will lose access to dedicated altimetry data.

NEMO was a joint government-industry effort to construct and launch an unclassified hyperspectral imaging system to support a broad range of commer-

[4]Presidential Decision Directive, National Science and Technology Council-2. 1994. "Convergence of the U.S.-Polar-Orbiting Operation Environmental Satellite Systems," The White House, Washington, D.C., May 5.

[5]See Appendix A for further background on the Navy's development of METOC systems.

[6]Funding data derived from recent DOD budget appropriations for FY98 through FY04.

BOX 4.1
Coriolis-WindSat: An Example of Interagency Cooperation in Satellite Development

Since 1985, satellites from the Defense Meteorological Satellite Program (DMSP) have incorporated the SSM/I (Special Sensing Microwave/Imager) radiometer, built by the Naval Research Laboratory (NRL), to measure speeds of sea-surface winds. Naval air operations also need information on wind direction, which can be obtained from the new, $70 million NRL-built WindSat microwave polarimetric radiometer now on the Coriolis satellite, launched successfully in January 2003.[1]

The Coriolis-WindSat mission spacecraft was built by the Spectrum-Astro Company; it was then modified to carry another payload, the Air Force Research Laboratory Solar Mass Ejection Imager (SMEI), and was space-qualified by the Naval Center for Space Technology. Both payloads successfully competed in the Department of Defense (DOD) Space Test Program (STP), which funded the Air Force launch and 1 year of operating costs.

Currently, the Air Force provides satellite command and control for Coriolis-Windsat, and routes the data through its Spacenet to NRL; from there the data go to the Fleet Numerical Meteorology and Oceanography Center. NRL and the Space and Naval Warfare Systems Command (SPAWAR) have also worked to ensure that the WindSat data can be downlinked directly to major combat ships, through the dedicated meteorology and oceanography (METOC) (SMQ-11) terminal, and made operationally useful. The sensor footprint is about 20 km in diameter and is conically scanned under the spacecraft at an angle of incidence of about 50 degrees, making a full (on-Earth) scan width of about 1700 km. Validation of WindSat-derived wind speed and direction measurements is planned, and will involve comparison with National Oceanic and Atmospheric Administration buoy measurements taken at sea level. SPAWAR currently acts as the manager for WindSat.

WindSat represents a risk-reduction program for the planned National Polar-orbiting Operational Environmental Satellite System (NPOESS) Conical Scanning Microwave Imager/Sounder (CMIS) sensor. Total on-orbit funding for Coriolis-WindSat was $224 million, of which $70 million came from the Navy METOC space program, $20 million from NPOESS, and $130 million from DOD's STP and the Air Force. The Navy anticipates that this work will result in CMIS being fielded onboard NPOESS.

---

[1]Michael A. Dornheim. 2003. "Coriolis Testing Earth, Space Weather Instruments," *Aviation Week and Space Technology*, January 13. See <http://www.ipo.noaa.gov/News/Archive/2003/jan/02/2003-01-13_Coriolis.pdf>. Accessed May 24, 2004.

cial and military needs. The Navy's stated interest was to demonstrate a satellite-based system to improve sensor coverage and information in the world's littorals. Government funding (approximately $70 million) was supplied by the Office of Naval Research (ONR) and the Defense Advanced Research Projects Agency

(DARPA). While this mission appears to have been strongly supported by the Navy, financial problems associated with the industrial partner appear to have led to the program's recent cancellation.[7]

GIFTS/IOMI was a recent Navy and NASA cofunded METOC satellite program. The GIFTS satellite was to test advanced technologies for measuring water vapor, wind, and chemical composition at high resolution over the Indian Ocean. Currently, the DOD relies solely on international partners for satellite-based meteorological data for this region. GIFTS appears to have received strong initial support from both NASA and the Navy, with an anticipated Navy investment of approximately $40 million from FY02 through FY04. However, the GIFTS program was recently cancelled after apparently suffering scheduling delays.

Other recent METOC programs of Navy interest include the NASA-funded Moderate Resolution Imaging Spectro-radiometer (MODIS) satellite, expected to provide 200 m resolution multispectral data. MODIS, with its 36 spectral bands across the visible and near-infrared, is a risk-reduction effort aimed to support the inclusion of a multispectral sensor on NOAA's planned National Polar-orbiting Operational Environmental Satellite System (NPOESS). The MODIS spectral bands are designed to enable monitoring of ocean color, phytoplankton, cloud properties, aerosols, and atmospheric water vapor in support of climate forecasting and global change research. In addition the Department of Energy (DOE) recently launched its Multispectral Thermal Imaging satellite, used to provide 20 m resolution global surface-temperature data. There are other examples of METOC programs of Navy interest as well. Several years ago, NASA flew a blue-green laser on the Space Shuttle to conduct research in ocean monitoring, and has flown experimental synthetic aperture radars to demonstrate ocean surveillance.

One drawback to NOAA's role as executive agent for environmental satellite development is that NOAA recently indicated to the Navy that active sensor systems (such as synthetic aperture radar or radar and laser altimeters) will not be placed on the next generation of national environmental monitoring satellites (Geostationary Operational Environmental Satellite (GOES) and NPOESS). This decision was made to support many of NOAA's requirements on imaging and sensor system performance. However, the decision leaves the Navy without a ready means to leverage its participation in NOAA programs to benefit the Navy's needs for active sensors. For instance, the Navy is currently supporting orbital altimeter systems (through Geosat and GFO), but these are experimental systems, not designed to provide the persistence and coverage necessary for use in tactical situations. In addition, the Navy's current reliance on GFO for altimetry data is

---

[7]Curtiss Davis. 2002. "Hyperspectral Imaging of the Littoral Battlespace," Overview Presentation, Coastal and Ocean Sensing Branch, Naval Research Laboratory. Available at <http://rsd-www.nrl.navy.mil/7230/pdf/7230_overview.pdf>. Accessed May 17, 2004.

scheduled to end with the planned termination of the GFO program in FY07. This loss of altimetry data concerns the committee, as it will leave the Navy (and the naval METOC community) without any altimetry data for at least several years after the GFO program's termination.

Finally, NOAA and the Navy have quite different missions and audiences for their products. NOAA focuses its resources on systems to provide weather predictions for the United States, while the Navy must be able to produce continually updated weather predictions for the entire navigable sea surface. These differences led NOAA to deploy the GOES satellites over the United States, leaving the Navy to rely on less timely data supplied from international partners to fill its global needs.

Sea Power 21

Three of the most pressing naval operational needs derived from Sea Power 21 are for systems enabling expeditionary warfare, countermine warfare, and shallow-water antisubmarine warfare in the littorals. Success in these warfare areas depends on local, timely environmental information. To help address naval needs, the Navy has developed a METOC Strategic Plan.[8] This plan lists three mission objectives relying on space-based environmental remote sensing to provide the following:

- Safe Operating Forces—Protect all assets.
- Optimized Warfighting Resources—Generate fiscal savings and increase military readiness through better forecasting of the global environment.
- Enhanced Warfighting Capabilities—Fully characterize the battlespace environment to the warfighter in terms that enable optimal employment of systems and platforms.[9]

The mission objectives Safe Operating Forces and Optimized Warfighting Resources tend to capture the largest share of the METOC community's resources. The primary products meeting these objectives are derived from the Defense Meteorological Satellite Program (DMSP), NOAA, and international partner satellite data, and synoptic and mesoscale atmospheric and oceanographic model output provided by the NMOC. Example products include forecasting services of NAVO's Optimum Aircraft Routing System (OPARS) and optimum track ship routing (OTSR). Environmental capabilities for the Enhanced Warfighting objective come largely from Polar Operational Environmental Satellite

[8]Naval Meteorology and Oceanography Command. 1997. *Strategic Plan*, Washington, D.C., May.

[9]Naval Meteorology and Oceanography Command. 1997. *Strategic Plan*, Washington, D.C., May, p. 3.

(POES), DMSP, and commercial sources, since they typically require higher spatial and temporal resolution than do the global forecasting models.

As described in the following subsections, these METOC mission objectives can be mapped into the Sea Power 21 pillars: Safe Operating Forces principally supports Sea Strike, specifically, carrier operations; Optimized Warfighting Resources supports Sea Basing, specifically, ocean routing; and Enhanced Warfighting Capabilities, the most challenging objective, supports tactical elements of both Sea Strike and Sea Shield. Appendix C presents additional detail regarding the dependency of Sea Power 21 capabilities on METOC products.

Safe Operating Forces—Sea Strike

The mission objective of providing Safe Operating Forces generally supports Sea Strike. It refers to providing atmospheric weather forecasts to get planes off carrier decks safely. NOAA currently supports this mission through Navy access to data taken by the GOES and POES systems. The geostationary network is globalized through international agreements with the European Space Agency's Meteorological Satellite (Meteosat), India's Indian National Satellite (INSAT), and Japan's Geostationary Meteorological Satellite (GMS). NOAA, as executive agent for environmental satellites, is responsible for managing international agreements.

Satellite data streams are received on carrier and other large-deck ships via the dedicated SMQ-11 environmental satellite receiver. POES and DMSP data are collected twice daily, at about 1 km spatial resolution. GOES data are broadcast every 30 minutes, at a resolution lower than that of POES or DMSP data. These data sets provide timely enough information to assist with tactical weather prediction near the U.S. mainland (because GOES is stationed to observe the United States). The military must then rely heavily on augmenting national capabilities when operating in foreign countries. While the Navy is able to access data through international partner agencies, there is often excessive latency for data received from these foreign assets, particularly from Meteosat and INSAT. The planned GIFTS/IOMI, recently cancelled, was being built by NASA with support from the Navy to satisfy NSS and civilian METOC needs over the Indian Ocean.

Optimized Warfighting Resources—Sea Basing

The mission objective Optimized Warfighting Resources is focused on the provision of OPARS and OTSR modeling services and generally supports Sea Basing logistics and scheduling needs. The current model, run at the NAVO facilities in Bay St. Louis, Mississippi, uses sea-surface wind and sea-surface temperature satellite data from DMSP and POES. The system can also be augmented by wave-height measurements supplied from the Navy's Geosat altimeter and wind-speed and direction measurements supplied by NRL's WindSat sensor onboard the Coriolis satellite. While recent advances in remote sensing have aided FNMOC's performance of these modeling activities, further improvements are still needed in the collection and archiving of detailed climatological information.

In addition, the Optimized Warfighting Resources objective includes the use of METOC information to assess optimal frequency use for tactical communications systems that cue appropriate ISR sensor systems, and to establish maps related to potential use areas for different classes of guided munitions. All of these resources rely critically on accurate and timely climatological information.

Issues for Safe Operating Forces and Optimized Warfighting Resources

Future needs for the two mission objectives Safe Operating Forces and Optimized Warfighting Resources include improved spatial resolution and timeliness of data access and update rates, as well as improvements in the overall support received from international partners. Most civil environmental satellite data have 0.8 km to 1 km spatial-resolution data sets and are supplied from passive sensors. There are plans to increase the resolution by about a factor of two in the next-generation GOES and NPOESS systems. The current data sets are generally satisfactory to enable the Navy's global and mesoscale forecasting and large-scale ocean modeling. Continuous improvement in satellite data for these two METOC missions should be adequately met in the future through civilian efforts, because the global environmental community is at least as interested as the Navy is in improved global and mesoscale atmospheric and ocean forecasting.

One concern in this regard, though, is the remaining reliance on foreign assets (Meteosat and INSAT in particular) and their associated large data latencies. As described above, NASA's GIFTS/IOMI satellite would provide a partial solution to this issue, but only if funding can be reallocated to the effort.

Enhanced Warfighting Capabilities, Including Sea Shield and Sea Strike

The METOC mission objective Enhanced Warfighting Capabilities includes tactical geospatial products—such as Special Tactical Oceanographic Information Charts (STOICS) and Special Annotated Imagery-Littoral (SAIL)—and model output from specialized tactical decision aids. These tactical products tend to be focused on the littoral regions, and they produce forecasting at finer spatial resolution and needing greater satellite repeat coverage than the general global forecasting services described above. For example, STOICS and SAIL typically rely on 1 m resolution data from DMSP, commercial multispectral imagers, and NSS sources. Timeliness, as discussed above, can be an issue for commercial as well as for foreign data that are often delivered in days or weeks, not tidal periods.

Examples of tactical decision aids are subsurface acoustic propagation predictions, ocean-surface front-and-eddy thermal analyses, and nowcasts of electromagnetic ducting conditions in the atmosphere's boundary layer.

Capability Gaps and Current Actions

The Navy currently relies on community partnerships to provide most of its satellite-based environmental remote sensing capabilities. The current METOC plan for developing satellite data to meet the needs of Sea Power 21 is to leverage R&D initiatives funded by NASA, NOAA, DOE, and other agencies, or to rely on commercially available data purchased by the National Geospatial-Intelligence Agency (NGA). This means that the Navy's future environmental data for enhanced warfighting will likely be limited to data that fulfill the spatial, temporal, and spectral requirements of the civilian climate change, civilian oceanographic, or commercial multispectral communities. While these data serve the needs of overall weather forecasting, there is currently no dedicated, space-based environmental support for conducting naval warfare in the littorals—including needs for expeditionary warfare, support for Special Operations Forces, shallow-water antisubmarine warfare, and countermine warfare.

Generally, these needs are first addressed through a substantive S&T program; however, the Navy has recently cut most of its funding support for advanced satellite-based METOC systems. Thus, virtually no Navy funding is available to use as leverage with other agency partners. For example, current funding does not appear to allow exploratory R&D with NASA and DOE satellites such as GIFTS or the Multispectral Thermal Imaging satellite. Limited Navy funds are, however, flowing into research for applications of future hyperspectral satellite systems and SBR.

The Navy has a long history of supporting METOC developments, but currently it does not appear to be expressing its unique interests through the various METOC partnership forums. This disconnect is hurting the Navy's efforts to satisfy its METOC needs. For example, hyperspectral systems for naval environmental littoral applications can be designed with useful signal-to-noise in the blue-water-penetrating regions of the spectrum; however, current national interests are focused on detecting man-made materials with associated sensitivity needs in the red and infrared spectral bands.

NOAA's current plans to limit future operations to passive systems leaves the Navy without its preferred partner when it comes to fielding active sensors, such as those with lasers or radars. No naval requirements for active systems (such as laser systems for bioluminescence detection, bathymetry, and so on) will be able to be transitioned onto NOAA satellites. This may necessitate the Navy's involvement with other partners or the possibility of the Navy's fielding its own satellites if it wants its needs for active sensing systems to be met.

Since the Navy is not currently developing any new environmental satellite programs of record (other than the current GFO program), any current Navy environmental satellite R&D efforts would have no defined transition path. As discussed in Chapter 3,[10] without a defined transition path programs funded through the Office of Naval Research, and through the Future Naval Capabilities (FNC) program in particular, are at a significant disadvantage when competing for advanced development S&T funds.

Recommendations Regarding Meteorology and Oceanography

Recommendation 4.5. The Department of the Navy should remain involved in developing and operating Navy-unique satellite systems. Thus, the Department of the Navy should reassess its meteorology and oceanography (METOC) remote sensing priorities. It is the view of this committee that these assessments should focus on the following:

- Ensuring strong support for the Geosat (Geodetic Satellite) Follow-on (GFO) program,
- Completion and launch of the Naval EarthMap Observer (NEMO) satellite, and
- Completion and launch of the Geosynchronous Imaging Fourier Transform Spectrometer/Indian Ocean METOC Imager (GIFTS/IOMI) satellite.

Recommendation 4.6. The Department of the Navy should pursue research and development of integrated active and passive microwave satellite sensors development programs with the National Aeronautics and Space Administration (NASA) and the National Oceanic and Atmospheric Administration (NOAA) to enable all-weather meteorology and oceanography sensing, along with measurements of trafficability, fog and visibility, and sea-ice mapping. The Navy should also continue to explore other research demonstrations, including active satellite systems and higher-resolution systems for hyperspectral imaging and sounding, atmospheric refractivity characterization and prediction, ocean color and biological constituents monitoring, and denied-area shallow-water bathymetry.

Recommendation 4.7. The Chief of Naval Research should modify the Office of Naval Research's technology transition rules to allow transition-oriented funds to support non-Navy (and non-DOD) meteorology and oceanography programs such as those fielded by NOAA and NASA.

---

[10]See the section entitled "Navy Space Support" in Chapter 3.

THEATER AND BALLISTIC MISSILE DEFENSE OF NAVAL FORCES

The broad category of theater and ballistic missile defense (TBMD) covers the availability of competent antiship cruise missile defense (ASCMD), overland cruise missile defense (OCMD), and theater ballistic missile defense capabilities. These capabilities will be essential if naval forces are to operate in littoral areas and execute the Navy's Sea Power 21 concepts of Sea Strike, Sea Shield, and Sea Basing. In addition, the current Marine Corps operational concepts—Operational Maneuver From the Sea (OMFTS) and Ship-to-Objective Maneuver (STOM)—envisage the use of light and highly mobile forces that are largely unencumbered by major air defense and artillery systems. These Marine Corps concepts of operations are thus based on the provision of TBMD, OCMD, air defense, and fire support by the Navy to support deployed ground forces.

Current threats to naval and joint forces operating in littoral areas stress the capabilities of current naval TBMD systems. Indications are that the future threats from hostile theater ballistic missiles (TBMs) and cruise missiles are likely to become more stressing as these systems become more and more widely available. Future cruise missiles are also likely to utilize features such as low-altitude, terrain-obscured flight paths; low radar cross-sections (RCSs); increased speed and agility; sensors that are resistant to electronic countermeasures; and precision terminal homing capabilities. (Note that U.S. forces are currently investigating all of these potential improvements to their own cruise missiles, so the eventual inclusion of the same improvements on adversary missiles is to be expected.)

As outlined in Chapter 2 of this report, the Sea Shield mission requires naval surface forces to provide responsive fire support for engaged forces ashore. The Sea Shield concept also requires that Navy ships be capable of providing overland air defense and TBMD. In turn, these mission requirements imply that naval surface forces must be able to operate safely in nearshore waters where their survival will be totally dependent on the availability of robust capabilities. A summary of the TBMD capabilities needed by Sea Power 21 is provided in Chapter 2 in Tables 2.2, 2.4, 2.6, and 2.8.

Cruise Missile Defense

Sea Strike and Sea Shield capabilities to defend against cruise missiles are provided with the SPY-1 air defense radar (aboard Aegis-class ships), the SPQ-9B surface search radar (scheduled to be replaced by the multifunction horizon search radar), and various versions of the semiactive Standard Missile (SM)-2 and shorter-range air defense missiles. The SM-2 missile engagement of incoming threat missiles has been limited to regions in which the threat missiles can be detected by the defensive radar (SPY-1) and illuminated by a fire control radar that can provide continuous engagement guidance for the SM-2. With the development of the Cooperative Engagement Capability (CEC), it is now possible to detect a threat missile with the radar on one platform and launch a missile from another platform to destroy the threat missile.

When naval forces operate in the open-ocean environment, this ensemble of cruise missile defense weapons and sensors provides the Navy with a reliable defense against antiship cruise missiles that do not have low RCS. Under such circumstances, the Navy's sensors can detect incoming missiles at ranges sufficient to provide a depth of fire that will allow either a shoot-shoot-shoot strategy or shoot-look-shoot strategy. However, naval forces are beginning to operate more often in littoral areas. In such environments, the ranges may be substantially degraded at which ship-based radars can detect low-RCS antiship cruise missiles following trajectories that obscure them from detection by defensive radar. In addition to the area defense concept based on the SM-2/SPY-1/multifunction horizon-search radar, the Navy's ASCMD capabilities include a number of shorter-range and point defense systems involving short-range missiles, guns, decoys, and electronic countermeasures to assist with defense against close-in missiles.

Overland cruise missile defense can present challenging situations when the line-of-sight paths that permit the detection and illumination of threat missiles are blocked by coastal hills or mountains. Thus, continued use of defensive missiles that are dependent on semiactive radar guidance (such as the SM-2) will become progressively more problematical as the Navy is asked to conduct more overland force-protection missions. An alternative to current defensive measures is to employ interceptor missiles that can be guided close enough to the threat missile so that the defensive missile's onboard sensor system can detect the threat missile and guide itself to closure.

Generally, OCMD is best accomplished using an elevated radar sensor with an airborne moving target indication (AMTI) capability; this sensor enables the detection and tracking of an incoming threat missile at extended ranges. However, unless the elevated platform carrying AMTI radar can also launch defensive missiles, its detection and tracking information must be relayed to a firing platform. Once a defensive missile has been launched, it must be provided with frequent guidance updates on the location of the threat missile. Current defensive missiles of the SM-2 family require update rates of 4 Hz. Thus, the maximum allowable latency for a communications system needed to enable closed-loop guidance for an SM-2-type missile is significantly less than 250 milliseconds.

Recognizing the critical importance of overland air defense and OCMD, the Navy has made extensive investments in the E-2C Radar Modernization Program (RMP). The RMP (which will incorporate space/time adaptive processing and the rotating ADS-18 phased-array antenna) is anticipated to provide the Navy with a competent overland AMTI radar capability. This RMP capability will also enable good performance even in the presence of overland background clutter. When combined with the new SM-5 missile, which will have multimode guidance and thus avoid the constraints of semiactive radar guidance, the Navy's capabilities for air and OCMD will be greatly enhanced.

While an excellent first step, the RMP may still have significant performance limitations. Further improvements in the overland AMTI performance of the E-2C RMP cannot be precluded. In particular, the E-2C normally operates well offshore and at low altitudes to enhance its survivability; this naturally limits some of the RMP's performance. Thus, cruise missiles launched from inland sites and programmed to fly low-altitude, terrain-obscured trajectories might still elude the RMP until the detection or clear line-of-sight range is too short to support an effective terminal defense of surface ships, supply depots, or engaged forces ashore.

If an AMTI SBR were available, the ability of a missile to avoid detection through the use of terrain obscuration would be minimized. Although an AMTI SBR would be the ideal solution to the problem of detecting cruise missiles that use terrain-obscured trajectories, the problem might also be alleviated to a degree if multiple, high-altitude unmanned aerial vehicles (UAVs) equipped with high-performance AMTI radars were available. To the committee's knowledge, no Navy acquisition program of record exists that is designed to produce a UAV-based AMTI capability.

If an AMTI SBR capability were to be developed, it would be necessary for the system architecture to provide guidance updates to the defensive missile with latencies of less than 250 milliseconds. This requirement would have a major impact on the design of the spacecraft and its associated communications networks. The AMTI SBR would need to have an onboard processor and a direct downlink to the platform controlling the flight of the defensive weapon. In a sense, the AMTI SBR concept would need to evolve from an ISR sensor into a tactical missile control radar.

An additional OCMD problem is that the performance of the E-2C RMP may be degraded by cruise missiles with extremely low nose-on RCS values. At any given radar-to-target range, a minimum detectable target strength always exists. The extremely low RCS values that are possible or that have been achieved for some cruise missiles are generally limited to nose-on aspects, in part because current defensive systems are designed to observe oncoming missiles nose-on. In situations (as is generally the case in air or OCMD scenarios) in which only nose-on detection is feasible, incoming missiles with extremely low RCS values may not be detectable until the range of first detection is so short that successful engagement is not feasible. Fortunately, it is significantly harder to configure a missile to have all-aspect low RCS values than it is to configure a missile with low nose-on RCS values. Thus, an overhead radar might have significant advantages over an airborne radar that is generally constrained to detect an incoming hostile missile from a nearly nose-on aspect.

There is a concern that naval representatives participating in discussions and cost-benefit studies with regard to the design of SBR are unfamiliar with the potentially important role that an AMTI SBR could play in overland air and OCMD engagements. The technology needed to support the development of an AMTI SBR system is generally acknowledged not to be mature enough at the present time. However, the committee was concerned that the Navy's R&D program did not appear to be directed toward the development of subsystems and components that could allow the use of a space sensor in a closed-loop weapons guidance mode.

In summary, the stressing problem of cruise missile defense may become even more challenging in the future. One of the most important future cruise missile defense sensors may prove to be an SBR with an AMTI capability. Current levels of technology support only a limited ability to deploy such a sensor system. To date, however, little or no R&D has been devoted to the technology that might provide the capabilities necessary for an AMTI SBR.

Theater Ballistic Missile Defense

The DOD has assigned primary responsibility for ballistic missile defense (BMD) to the Missile Defense Agency (MDA). However, because of the importance of theater ballistic missile defense, the Navy has devoted significant resources to the development of responsive defensive systems. In general, a fully deployed theater ballistic missile defense system likely will be based on integrating the capabilities provided by the following:

- Patriot Advanced Capability-3,
- Medium Extended Air Defense System,
- Theater High Altitude Area Defense System,
- Airborne Laser (ABL),
- Navy Area Defense (NAD) system (or equivalent), and
- Navy Theater Wide (NTW) defense system (or equivalent).

To operate effectively, many of these systems will need to be cued by space-based national assets, such as the Defense Support Program (DSP) and, in the future, by the Space-Based Infrared System-High (SBIRS-H) and networked in a battle management command, control, and communications system.

The development a robust theater missile defense capability demands technological advances in a number of areas in addition to that of space sensor performance. These include the need for improved long- and short-range surveillance sensors, guidance and control, propulsion, automated target-tracking technology, data processing, and lethal intercept techniques (kinetic kill, explosive charge, and the like). These systems will also rely on command, control, and communications networking for various purposes, including the provision of cueing and links of naval components to National Technical Means (NTM) and future BMD systems such as ABL. Thus, the ability to engage an attacking ballistic missile successfully will depend on the availability of effective hit-to-kill interceptors, multispectral seekers, and cooperative over-the-horizon surveillance and fire control. It will also depend on the ability to detect, evaluate, and overcome penetration aids and other countermeasures to theater missile defense.

Current national, space-based sensor capabilities that have relevance to ballistic missile defense include these:

- The DSP, and
- Various signals intelligence (SIGINT) collection programs and their associated Tactical Receive Applications Program/Tactical Receive Equipment information dissemination systems.

Although these programs provide useful cueing, they are not structured as low-latency systems that can be used to provide real-time, closed-loop weapons guidance data. For instance, DSP satellites detect missiles in their ascent phase as soon as the missiles have risen above the cloud deck. If DSP can provide a missile's velocity and direction of flight at burnout, then the missile's trajectory and probable intended impact area can be inferred. Unfortunately, the data rate of the DSP sensors is relatively slow, being constrained by the spin rate of the space vehicle—DSP is a spin-stabilized spacecraft, rotating a few times per minute, using the satellite's spin motion to scan an array of infrared detectors, operating in the short-wave infrared range, to detect the emissions from rocket plumes during the boost phase of a launch. Thus, it can take a significant fraction of a minute for the DSP system to declare a detection. Since most ballistic missiles reach burnout in less than 3 minutes, the detection process consumes a considerable fraction of the time available during the ascent stage.

A single DSP satellite gives limited geolocation data relative to the launch site and, because of the multisecond frame rate, the resulting track has a large propagation uncertainty. However, if two DSP satellites can view a launch simultaneously (binocular DSP) and their track information linked better, launch-point geolocation can be achieved, and the azimuth of the missile's trajectory can be better predicted. While such data do not provide precise trajectory information, they certainly limit the volume that must be searched by the defensive radar. This cueing allows the radar to focus its radiated energy into a significantly narrower angular cone and thus increases the initial detection range and accuracy of the radar.

DSP is scheduled to be replaced by SBIRS-H. SBIRS-H is designed to track missiles during powered flight and to provide much higher precision tracking information than DSP does. This will be possible because the spacecraft will be a three-axis-stabilized vehicle, and the SBIRS-H optical system is based on a large, high-sensitivity, focal-plane array that can be adaptively scanned at high frame rates. This capability is planned to allow SBIRS-H to detect low-intensity rocket plumes, track the detected plumes, and provide greatly improved missile trajectory measurements. In addition, SBIRS-H is designed to have much greater onboard signal and data processing capability than that of DSP and will also have the ability to directly crosslink to theater platforms and thus eliminate the need for overseas ground data entry points. As a result, SBIRS-H, if deployed, could provide excellent tactical warning and attack assessment and cueing data directly to forces in the field as well as to the National Command Authority. While this system has been in development for many years, its delivery date is still uncertain. This uncertainty is a matter of concern, since many of the Navy's (and MDA's) needs for space-based early warning are based on use of SBIRS-H.

The Space-Based Infrared System-Low (SBIRS-L) satellite system, while not yet approved for full-scale development, is conceptually designed to operate in the visible and long-wavelength infrared, 8 to 14 μm, looking at targets against the cold space background. The task of SBIRS-L would be to provide the midcourse tracking of ballistic missiles in flight and to hand off the target(s) to a midcourse or terminal defense system. With its multispectral sensors it could, in principle, provide some midcourse discrimination of warheads and decoys.

As a consequence of the decision not to proceed with the Future Early Warning System, the SBIRS-H acquisition design parameters were changed to include a capability to detect intermediate-range ballistic missiles and short-range ballistic missiles. To achieve this capability, a higher scan rate, increased sensitivity, a new detection band at 4.3 μm, and a two-dimensional focal plane array were added to SBIRS-H.

The Navy originally envisioned reliance on the Aegis weapon system, employing variants of the SM-2, to provide an in-theater (at sea) capability to engage ballistic missiles within the atmosphere (this concept is the current version of the NAD program). An advanced variant of the standard missile (the SM-3) would provide a capability to engage ballistic missiles at longer ranges outside the atmosphere (the design mission of the NTW program). Under the Sea Shield concept, forward-deployed naval forces are envisaged as being capable of making this contribution during the developing phases of a conflict, when they would be called upon to protect threatened nations and arriving joint forces against attacks by ballistic missiles.

The NTW system concept is not as mature as the NAD system concept. The Achilles heel of the NTW program is the Aegis SPY-1 radar, which is an excellent air defense radar but a marginal radar for the full range of NTW mission requirements. For ascent-phase engagements, which may be an important role for NTW, the large RCS of theater ballistic missile booster rockets may support adequate use of the SM-3 interceptor. However, even in an ascent-phase engagement, the SPY-1 radar would probably need to be cued by an external sensor so that all its available beam energy could be focused on the incoming missile. The only space-based sensor currently available to cue the SPY-1 is the DSP satellite, with the deficiencies described above. For more reliable cueing to support ascent-phase engagement, an overhead sensor with the capabilities postulated for SBIRS-H appears to be needed.

Many situations may require a midcourse-engagement capability rather than an ascent-phase engagement capability. For midcourse engagements, the hit-to-kill warhead of the defensive missile will require sophisticated capabilities for both guidance and decoy discrimination. In such a situation, the SPY-1 will require external cueing by an overhead sensor with the capabilities postulated for SBIRS-L. In principle, alternate space-based sensors might be deployed that would support the decoy discrimination function. To the best of the committee's knowledge, no significant Navy programmatic effort is under way to develop such alternate capabilities. In addition, neither SBIRS-H nor SBIRS-L is progressing at a pace that inspires confidence that it will be available in a time frame reasonable for assuming the cueing function for the NTW system.

The committee recognizes that for R&D in support of new ballistic missile defense capabilities, the Navy cannot proceed autonomously. Current DOD directives stipulate that all missile defense R&D programs be coordinated with and supported by the MDA. Although there is substantial Navy representation within the MDA, there is little indication of what more Navy representatives could do that is not part of MDA's response to the problems of SBIRS. The SBIRS program has been beset with unresolved technological challenges and problems of cost growth and requirements creep. While the Navy may be able to field a marine-based ascent-phase BMD capability, a true ability to implement the Sea Shield concept will not occur until appropriate space-based sensors become available. There is nothing that the Navy can do about this situation, because the authority and responsibility for the development of this capability rest with the MDA.

One means to augment MDA capabilities for BMD would be through the development of an AMTI SBR. However, there are no current NSS plans to develop a space-based persistent AMTI radar similar in capability to that provided by current E-2C and Airborne Warning and Control System (AWACS) radars. Such an NSS capability, if it were developed and fielded, would dramatically expand, to global dimensions, the range for detecting and targeting airborne threats from ranges currently provided by UAVs or piloted aircraft (hundreds of square miles). While creating such a system would represent a stretch with current NSS sensor technology, this is an area in need of further study by a concerted S&T program with the objective of identifying future NSS ISR potential in this area. A nearer-term NSS opportunity to address the missile defense mission could also involve a persistent, multiple-look-angle NSS infrared detection capability. Individual satellite sensors might each be similar in capability to that available with SBIRS-H, and multiple satellites with a view of all points of interest on Earth could permit detection and tracking of theater ballistic missiles that must be defended against by Sea Shield.

Recommendations Regarding Theater and Ballistic Missile Defense

Recommendation 4.8. The Navy should continue its aggressive support of the E-2C aircraft Radar Modernization program so that a fleetwide capability can be achieved as soon as feasible.

Recommendation 4.9. The Department of the Navy should begin operational analysis of the cost, benefits, and requirements of a cruise and ballistic missile defense system based on a multimode missile and an airborne moving target indication (AMTI) space-based radar (SBR) system. The Department of the Navy should invest in a focused science and technology program to resolve the issues that currently render an AMTI SBR infeasible.

SPACE-BASED COMMUNICATIONS

Space-based communications are embedded in almost every portion of Sea Power 21. In order to support the President's desire that the U.S. military "be ready to strike at a moment's notice in any dark corner of the world,"[11] there must be continuously available, low-latency, high-assurance global communications between sensor and processor, strategist and planner, and commander and tactical forces. Most beyond-line-of-sight (BLOS) links to deployed tactical units currently use space-based communications systems. The role of space-based communications in supporting key elements of Sea Power 21 is discussed in the next subsection, followed by a description of the Navy's role in providing these capabilities, and information on the gaps between the capabilities of space-based communications systems and the needs of Sea Power 21. A brief discussion of current and planned DOD space-based communications programs is provided in Appendix D. This section ends with consideration of specific findings and recommendations for closing these gaps.

Background on Satellite Communications

To provide a common level of understanding regarding communications, this section discusses some basic elements of the physics of electromagnetic waves and the regulatory environment for radio spectrum use. Physical laws constrain design and use issues, such as antenna size and placement and the attenuation and detectability of signals affected by adverse weather, while regulations specify when and where in the world various portions of the spectrum may be used, and who may use them. Figure 4.1 depicts current allocations of satellite radio frequencies for U.S. government and commercial services. As shown, a broad range of

---

[11]President George W. Bush. 2002. Address to the United States Military Academy graduating cadets, West Point, New York, June 1.

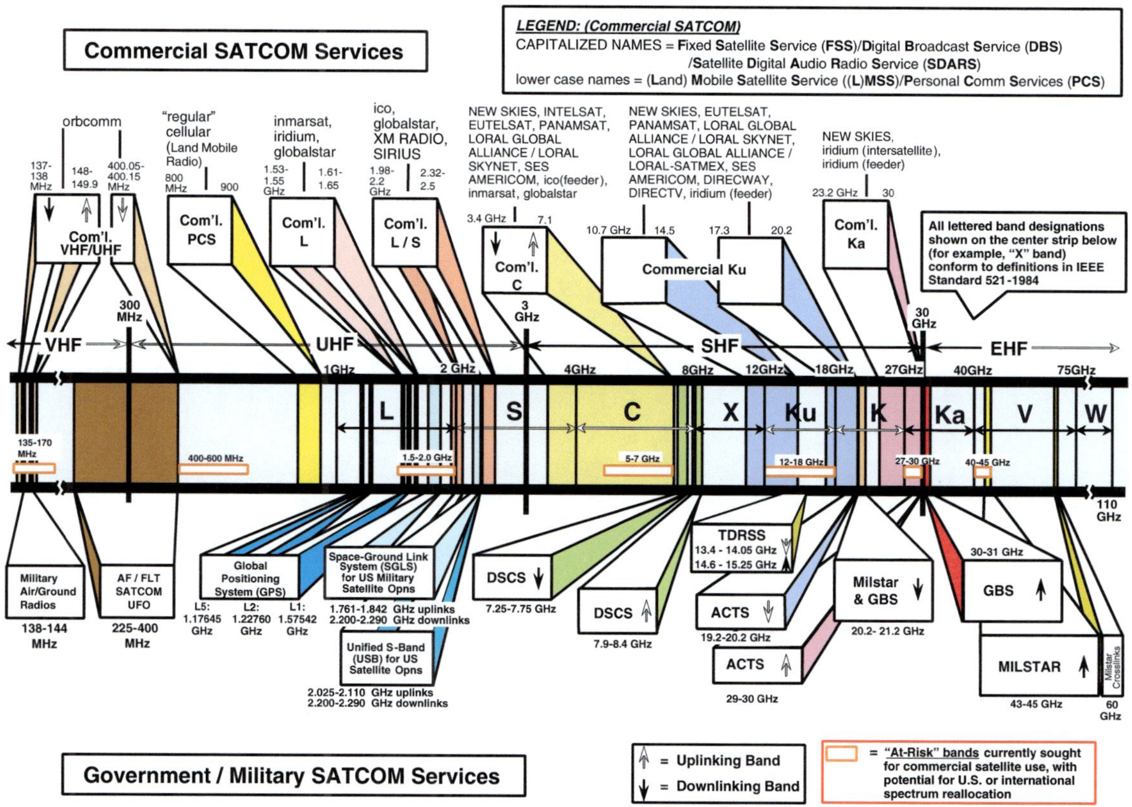

FIGURE 4.1 Commercial and U.S. government satellite frequency allocations. NOTE: A list of acronyms is provided in Appendix G.
SOURCE: Courtesy of Dennis L. Mauney, Information Technology and Applications Corporation (ITAC), Reston, Va.

frequencies is currently employed for satellite communications, ranging from very high frequency (VHF) bands at the low end, through ultrahigh frequency (UHF) and super high frequency (SHF) bands, up through extremely high frequency (EHF) bands at the current high end. Certain specific subbands, such as the C, Ku, and Ka bands, are also well-known designations in their own right.

Laid out across these spectrum bands are both commercial and government satellite communications systems. For example, Inmarsat and Iridium are commercial services in the L band (1.5 and 1.6 GHz) while Global Broadcast Service (GBS) is a government system in the Ka band. In the future, the Mobile User Objective System (MUOS) will occupy the bands labeled UHF Follow-on (UFO) system, and the Wideband Gapfiller System (WGS) will occupy the X, C, and Ka bands.

As a general rule, higher-frequency systems provide more user capacity (bandwidth) than systems lower in the radio-frequency (RF) spectrum. For example, the Advanced Communications Technology Satellite (ACTS) system in the Ka band provides roughly 1 Gb/s to a terminal, in contrast with Inmarsat, which provides on the order of 32 kb/s in the L band. Although to some extent this is simply because there is more RF spectrum available in the high bands, overall it is as much a matter of technological and regulatory history as anything else. For instance, reasonably high bandwidth satellite services could be provided even in the rather limited UHF bands if the military Services were willing to dispense with the decades-old legacy of subdividing the UHF band into fixed 25 kHz channels.

Figure 4.2 provides a useful sketch of the user terminals currently employed by the Services to access satellite communications. Note that the new MUOS program will be significantly driven by the preexisting community of 82,000 UHF satellite communications (SATCOM) terminals for handheld and vehicle-mounted systems; whereas the X- and Ka-band systems have, in general, been designed for far fewer, but much larger, user terminals, thus enabling faster and easier Service-wide terminal upgrades.

In general, high-capacity satellite links require larger antennas than those needed for low-capacity links. This difference has obvious platform implications for the Navy and Marine Corps. While big-dish antennas 7 to 9 ft in diameter may easily fit on large, surface ships, they are infeasible for the advanced amphibious assault vehicle of the Marine Corps or for submarines. Thus, large-deck ships are far more likely to enjoy direct, high-capacity satellite links than are smaller platforms or dismounted units.[12] Most high-capacity satellite systems also rely on

---

[12]To this end, it may be desirable to create a FORCEnet network architecture with large-deck or otherwise advantaged platforms as satellite downlink and uplink "hubs" that relay packets to other platforms via a variety of near-Earth networks, such as provided by the Joint Tactical Radio System (JTRS) Wideband Networking Waveform. Such a system is typically described as a hybrid communications system.

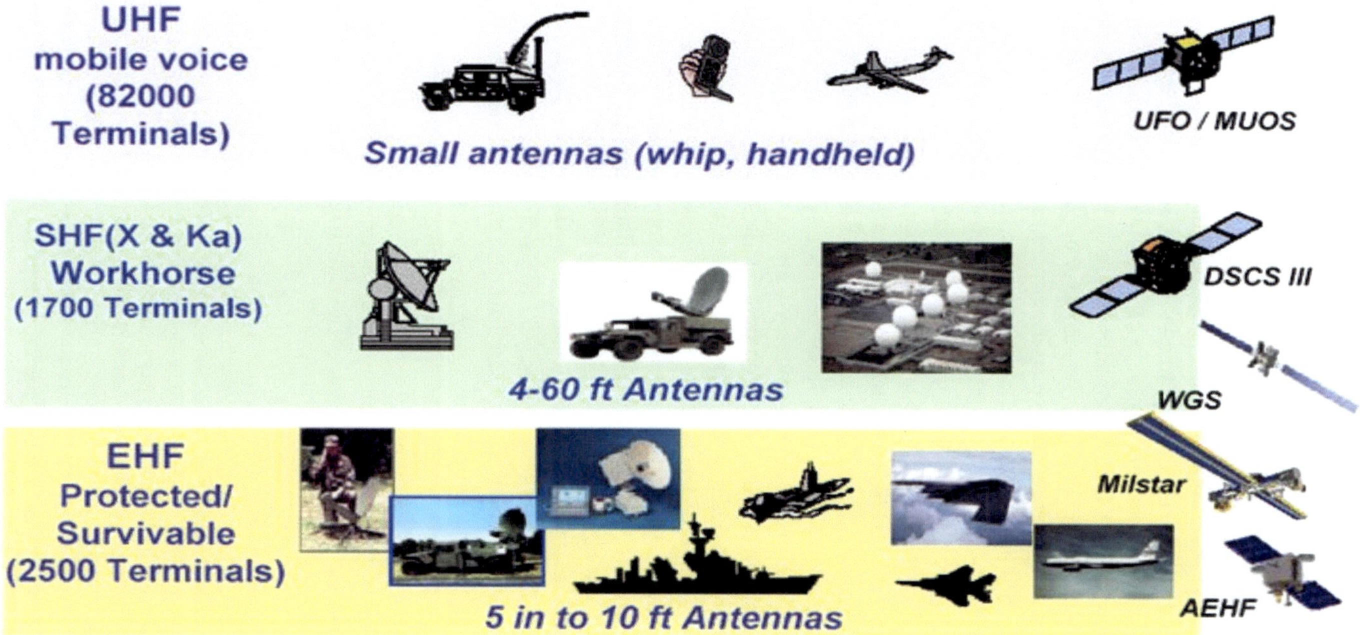

FIGURE 4.2 Types and number of terminals used by the Services to access current (and planned) communications satellites. NOTE: A list of acronyms is provided in Appendix G. SOURCE: Christine M. Anderson, Program Director, Military Satellite Communications Joint Program Office. 2002. "Transformational Communications," slide 6, presentation to Ground System Architectures Workshop, El Segundo, Calif., March 14.

nongeostationary satellite constellations. Hence, the associated antennas must be continuously steered not only to account for the motion of the platform, but also to account for the motion of the satellite.

Another difficult area for naval communications is the need to keep a clear line of sight between the antenna and the satellite. In many shipboard antenna placements, the superstructure may block this line of sight as the platform maneuvers. To compensate, the platform may need multiple, linked antennas positioned so that at least one antenna always has a clear view of the satellite. Navy platforms may need as many as four linked antennas in order to achieve acceptable (greater than 99 percent) availability, or they may need to have the antenna systems moved to higher positions relative to the superstructure.

Optical communications systems (called lasercom) are conceptually similar to highly directional, high-band RF systems—though laser transmitters and receivers are physically quite different from radio equipment. Lasercom is capable of very high bandwidth capacity, up to tens of gigabits per second to a terminal and, as envisioned by the DOD Transformational Communications Architecture (TCA) program, will provide the high-capacity data links of the future. Numerous studies have concluded that optical links between seaborne and satellite platforms are feasible; however, optical communications can be severely impacted by weather, turbulence, and other obscurants. Studies within the continental United States have concluded that a single optical link's Earth-to-space availability is on the order of 55 percent at best (as measured in "sunny" Roswell, New Mexico) and is driven to a large extent by relatively long term phenomena such as cloud cover.[13] Even quite complex schemes, with available optical bandwidth dynamically derived from cloud-scatter pulse-dispersion models, required three sites broadly scattered across the continental United States to achieve greater than 90 percent availability for a relatively modest 100 Mb/s of optical link.[14] Direct optical links from sea or Earth platforms to satellites will therefore be difficult to implement for the mobile tactical user.[15]

While many current efforts (such as TCA) are aimed at providing direct big-bandwidth links to mobile users, other alternatives do exist. Among these are information compression systems as simple as text messaging. During Operation Iraqi Freedom, the Services all made extensive use of text-messaging capabilities

---

[13]Sabino Piazzolla and Stephan Slobin. 2002. "Statistics of Link Blockage Due to Cloud Cover for Free-Space Optical Communications Using NCDC Surface Weather Observation Data," *Proceedings of the Society of Photo-Optical Instrumentation Engineers*, Vol. 4635, pp. 138-149.

[14]Daniel V. Hahn, Clinton L. Edwards, and Donald D. Duncan. 2002. "Adaptive Compensation of Atmospheric Effects with a High-Resolution Micro-Machined Deformable Mirror," *Proceedings of the Society of Photo-Optical Instrumentation Engineers*, Vol. 4821, pp. 320-331.

[15]However, the Defense Advanced Research Projects Agency Optical RF Link Experiment may provide a significant mitigation by implementing a hybrid radio-frequency/optical link in a single aperture. This system is intended to employ radio frequency at all times, but also to take advantage of optical connectivity when available.

to create tactical chat rooms. The low bandwidth inherent in these systems allowed for greater communications and interaction between combat elements even when bandwidth resources became limited. Another alternative to direct high-bandwidth links is the employment of hybrid hub-and-spoke architectures. The hybrid systems proposed typically rely on a single, dedicated high-bandwidth satellite link to a single server (often envisioned as a capital ship or a high-altitude unmanned air vehicle) that relays the information via high-bandwidth line-of-sight links to the mobile users.

The Role of Space-Based Communications in Sea Power 21

Communications constitute a critical function for each of the pillars of Sea Power 21, the Navy's strategy for implementing naval transformation. The allocation of communications functions and the scaling of performance parameters between space, airborne, and terrestrial systems are not clearly delineated by the Navy in descriptions of Sea Strike, Sea Shield, Sea Basing, and FORCEnet.

In order to support modern military and naval operations, there must be continuously available, low-latency, high-assurance communications between sensors and processors, strategists and planners, and commander and tactical forces worldwide. Most BLOS links to deployed tactical units are now transmitted via space-based communications systems, and as bandwidth demand rises, so will the Navy's dependency on space-based communications assets. For example, in order to conduct Sea Strike operations, space communications linking of information from sensors, to analysts, to decision makers, to the warfighters will be necessary. Sea Shield relies critically on space-based communications to provide individual and fused threat information quickly and to cue theater and strategic missile defense assets. Sea Basing cannot function without the high bandwidth necessary to link commanders to tactical units throughout the theater, and to link the chain of command for rapid strategic and tactical planning and decision making. Finally, the FORCEnet concept requires the flexible, low-latency, worldwide communications that are only enabled by space-based communications capabilities. Each of the major concepts of Sea Power 21 is thus critically dependent on space-based communications, and a review of the specific elements within Sea Power 21 reveals that most of them carry fundamental dependence on space-based communications systems. These Sea Power 21 capability dependencies on communications are summarized in Chapter 2, in Tables 2.2, 2.4, 2.6, and 2.8.

Space-Based Communications Support for Sea Strike

Sea Strike operations will involve the dynamic application of strike, naval fire support, ship-to-objective maneuver, and strategic deterrence to deliver devastating power and accuracy in future campaigns. The primary function of space-based communications in Sea Strike is to provide assured, all-weather, reliable, timely, and accurate communications from ISR sensors, analysis, exploitation, and data fusion processors and centers to the diverse command, control, communications, computers, intelligence, surveillance, and reconnaissance (C4ISR) systems linking strike weapons, platforms, and infrastructure. Some of the more important space-based communications links in strike operations include the following:

- Defining and coordinating missions with other command elements;
- Requesting appropriate support, including logistics and search and rescue;
- Tasking and relaying data requests from nonorganic sensors;
- Providing communications of information from space-based, airborne, and ground-based sensors to processing and interpretation centers;
- Supporting data fusion operations by communicating raw and processed information to interpretation, data fusion, information analysis, target identification, and target selection elements;
- Providing selected communication of strategic and tactical targeting information to weapons delivery systems; and
- Relaying post-strike information for damage assessment and interpretation.

In addition, space-based communications have an increasingly important role in recent conflicts, supporting Special Operations Forces and Marine Corps operations in littorals as well as supporting forced entry and other small, mobile ground force operations. These operations have heavy dependency on the requirement for high-availability, high-assurance, low-probability-of-intercept, low-latency, all-weather communications. In these operations, antenna size, portability, and ease of setup and disassembly are critical. Both voice and digital data are required, as is connectivity for weather, space- and ground-based intelligence information, and integrated near-real-time threat assessment. This capability, with increased numbers of users being supported, was a top priority identified recently by the Commander, U.S. Central Command.[16]

Providing these space-based communications capabilities across all of the naval elements in a manner that is both affordable and places minimal constraints on the operations of various elements is a goal that has not been met. Bandwidths are oversubscribed, communications links have marginal availability at times, and new capabilities in development (e.g., streaming video) cannot now be easily supported from space. Some of these capabilities will be provided by the TCA

[16]Written response by General John Abizaid, USA, Commander, Central Command, during his congressional confirmation hearing, June 24, 2003. Available at <http://armed-services.senate.gov/statemnt/2003/June/Abizaid.pdf>. Accessed May 17, 2004.

program, the Joint Tactical Radio System, and other elements currently under development by the Air Force and other Services and agencies. However, because of low levels of space-based communications S&T funding, the naval forces do not appear to be advancing options to potentially address the complex issue of achieving the Navy's communications goals.

The only new ideas presented to the committee to address these issues were presented by the Naval Center for Space Technology (NCST). The NCST concept of a geosynchronous, ultrawideband payload addressed two of the critical naval needs: scalability to high bandwidths and small apertures. Furthermore, this concept proposed that spot beams and bandwidth could be directly controlled by the Joint Task Force Commander. Unfortunately the S&T funding to evaluate this concept fully does not start until FY05 and is provided at a low level until FY07. Thus, by the time the study's results are in hand, most of the TCA program's early trade-offs will be completed. This eventuality again points to the Navy's need for continuous and sustained S&T support, since time often will not allow for the Navy to initiate an S&T activity and get results soon enough to support other Office of the Secretary of Defense (OSD) or joint decisions.

The lack of an integrated Navy communications strategy is reflected in the separation of conventional communications links from space data links, resulting in limited distribution of important data. For example, the tactical airborne reconnaissance pod system (TARPS) (for F-18 E/F aircraft) provides a large-area ISR capability that is communicated to ships by the common data link, a line-of-sight communications system that has a bandwidth of 274 Mb/s. However, current Navy ships typically have at most 8 Mb/s of satellite communications (BLOS) capacity to communicate these data off the receiving ship; hence, only meager snippets of information can be transferred to other users. This is contrary to the seamless, wideband precept of the Global Information Grid (GIG) and inhibits implementation of the TPPU concept.

The next generation of many types of space-based ISR sensors is being designed to collect data at many gigabits per second. Hence, without significant improvements to maritime antennas, naval forces will potentially be deprived of information that has been collected but cannot be distributed to naval platforms. This capability gap can only be overcome through careful communications planning and implementation of systems with improved bandwidth, latency, and availability.

Space-Based Communications Support for Sea Shield

To maintain littoral superiority for naval and joint force components, communications resources must be able to support protection against conventional and unconventional (i.e., chemical, biological, radiological, nuclear, or explosive) threats from special operations and terrorist forces as well as threats that might be mounted by more conventional enemy ground forces. Information from space-, ground-, and sea-based and airborne ISR resources must be communicated to neutralize near-horizon and over-the-horizon threats. Such information is needed to enable deep-ocean and littoral operations by supporting self-defense against or neutralization of undersea threats (including submarines, mines, submerged barriers, and obstacles) and to provide defense over land and over sea against theater air and ballistic missile threats. The Sea Shield mission requires that naval forces establish air control against hostile aircraft and be capable of mounting a successful defense against cruise and ballistic missile attack, both in naval operating areas and as far inland as practicable. The Sea Shield mission also requires naval forces to assure their own survival (afloat or ashore) and the survival of their associated air and surface logistic forces.

Sea Shield operations currently depend on terrestrial line-of-sight communications for battle group defensive capability based on the CEC,[17] while depending on space-based communications for general support in detecting, identifying, and neutralizing OTH threats, deep-water mine fields and for linking and communications between theater and national sensor systems and command structures. Expanding the CEC and other new capabilities, such as wide-area distributed undersea operations, will also require low-latency, high-assurance communications. Space communications will be an essential link in establishing extended capability, but with respect to Sea Strike, naval S&T has not yet been focused on the these issues. The NCST recently introduced two concepts to examine options for the distributed sensing: one is an Internet Protocol (IP) net-distributed expeditionary sensor, and the other is on-demand conflict support via tactical microsatellites.[18] A proof-of-principle example of the latter, supported by the OSD Office of Transformation, has a focus on rapid support from space but not on associated communications needs.

In the emerging area of theater missile defense, space-based communications will provide the capability to link national, threat-sensing systems such as the Space Surveillance and Tracking System to command-and-control and countermissile defensive systems. Space links will be vital in linking information from BLOS sensing systems in many theater or strategic missile defense scenarios, including geolocation and computation from space-based assets of the trajectory of inbound tactical and cruise missile threats.

Finally, communications links to SBR will provide large amounts of data over even larger areas than are covered by the TARPS capability described above. Further, SBR may also have a potential role in the target identification of low-observable cruise missiles, should AMTI functionality be included in the system. For this to be effective, very low latency is essential, and current concepts for GIG transport services do not appear to support such demands.

---

[17]See the associated discussion in the preceding major section.

[18]Peter G. Wilhelm, Director, Naval Center for Space Technology, "Space S&T Initiatives Supporting the Navy's Role in Space," presentation to the committee, June 27, 2003.

Thus, space-based communications will provide essential capability in support of Sea Shield. They will provide OTH sensor and command information to the fleet and must provide high-bandwidth, assured, low-latency communications.

Space-Based Communications Support for Sea Basing

Sea Basing provides the operational platform capability from which the Navy will project both offensive strike (i.e., Sea Strike) and defensive protection (i.e., Sea Shield) as well as supporting the Marine Corps and joint forces as appropriate to the missions. Sea Basing is accomplished by using the fleet assets with platform, logistics, and communications improvements, but without extensive use of existing port facilities and logistics ashore. A significant space communications demand arises when the sea basing includes the forward-deployed command center. Joint forces' command center communications needs are particularly stressing. For example, during Operation Iraqi Freedom, estimates of peak communications to support operations of the joint forces were over 750 Mb/s.[19] This level is nearly 75 times the planned capacity for any Navy ship (prior to FY07).[20] With data and bandwidth usage roughly doubling each year across the DOD, the next decade could bring the Navy's deployed communications capacity needs into the tens of gigabits per second for each naval platform; this would aggregate to terabits per second for total fleet bandwidth capacity needs.

Sea basing of the joint operations command center provides the final example in which the evolving concepts of Sea Power 21 will cause explosive growth in space communications needs. On the sea base, large-scale planning will require access to large, globally dispersed databases, and since most data will be collected away from the sea base, rapid agile planning on the sea base will require extremely large bandwidth communications in order to collect and assess planning information, to implement high-quality video conferencing, to access near-real-time sensor information (e.g., SBR), and to implement the TPPU concept of linked worldwide DOD information data structures.

Space-Based Communications Support for FORCEnet

FORCEnet is the networked communications and information collection, fusion, and processing capability required to implement the command, control,

---

[19]Lt Gen T. Michael Moseley, USAF, Commander, Central Air Forces. 2003. *Operation Iraqi Freedom—By the Numbers*, Shaw Air Force Base, S.C., April 30, p. 12.

[20]By FY07, the Navy is planning for large-deck ships to have bandwidths of 10.5 Mb/s at most. See the subsection below entitled "Space-Based Communications Capability Gaps and Issues" for further detail.

and communications functions of Sea Power 21. FORCEnet focuses on the gathering, processing, transportation, and presentation of information in support of the scope of the Sea Power 21 vision. It is planned to be the integrator and enabler for the three pillars Sea Strike, Sea Shield, and Sea Basing. It will rely on programs such as the GIG and TCA to implement information transmission worldwide, and is focused on providing the communications infrastructure, network protection, and information-assurance functions internal to the network. In addition, FORCEnet provides an integrated common operational and tactical database. FORCEnet core capability is based on the implementation of an IP-based, Internet-like protocol and on the adoption, where possible, of commercial standards for communications. Force projection and defense from forward-deployed naval platforms depends on the efficient networking of naval, national, and force nodes involved in all aspects of information production, command responsibility, and control authority.

The dependence of FORCEnet on space-based communications is implicit in the domain of FORCEnet functionality. FORCEnet has not developed, as far as the committee can discern, a systems-engineered view of the connectivity and capability required from space-based communications systems. This is a major shortcoming that needs to be addressed so that the Navy can define and defend its requirements to the DOD community, can plan and allocate resources, and can articulate S&T needs.

Some of the space and terrestrial communications needs that FORCEnet is built upon are as follows:

- A global communications capability, with sufficient diversity to ensure accurate and timely information communications in order to enable unencumbered naval operations;
- Access to all categories of tactical and strategic information, both digital and voice;
- Information availability and assurance commensurate with data type, category, and priority;
- Diverse, robust, and redundant communications pathways to overcome communications loss owing to threats, antenna blockages, or weather effects;
- Information types and volumes tailored to the needs of sending and receiving data among systems, platforms, and users;
- Timely delivery of information;
- Dynamic, programmable allocation of bandwidth;
- The ability to grow communications capacity efficiently and gracefully over time;
- Maintaining of compatibility with legacy systems, including EHF, SHF, GBS, and UHF; and
- Support for all naval organizational needs, including Navy and Marine Corps platform needs (sea/undersurface/air/land/space), dismounted operational

Navy and Marine Corps unit needs (sea/land), and Navy and Marine Corps needs in the areas of command, control, administration, organization, and logistics.

As discussed above, major shortfalls already exist in the scale of the bandwidth that supports the Navy, but the gaps will grow as other organizations supported by Sea Shield and Sea Basing increase their information bandwidth demands on Navy communications systems. In particular, these demands will involve the current NGA migration toward greatly increased bandwidth sensors outputting high-quality, high-definition-television information and streams of UAV video; powerful new video teleconferencing tools being deployed to improve planning and coordination (some of which require up to 6 Mb/s of capacity); and the Distributed Common Ground System, which is being developed to transmit large sensor data sets over high-rate (a few gigabits per second) links. Lastly, the DOD's implementation of the TPPU concept has the potential to greatly increase the demands for space-based information bandwidth. Under TPPU, large numbers of users will be accessing relatively unprocessed data directly collected from sensors. For example, the large amounts of data that the TARPS E/F collects under this model should be posted for all other potential users to access. While there are many ways to implement the solutions (e.g., by shipboard server farm storage, terrestrial data warehousing, and so on), they all involve very high bandwidth space-based communications among warfighting platforms and storage locations. So, worldwide high-bandwidth, high-availability space-based communications are essential for implementation of the TPPU concept.

Space-Based Communications Capability Gaps and Issues

Bandwidth Needs

Future bandwidth available for naval forces can be understood in two ways—as an aggregate available bandwidth for a theater and as the maximal available bandwidth for a given platform. The aggregate is limited by the Navy's upper bound on allocations of bandwidth from satellite systems. A platform's bandwidths are typically limited (1) by the installed apertures and terminals on the platform, assuming that there is enough theater bandwidth available to service that platform; and (2) by the electronic capability to multiplex and handle the various data types flowing to and from the space segment.

Figure 4.3 shows the estimated amount of wideband communications bandwidth available to the afloat Navy during the spring of 2003. It shows that the total U.S. fleet had available approximately 192 Mb/s of bandwidth. Note that the majority of the indicated commercial and DOD resourced bandwidth was provided by unprotected communications (Inmarsat, Defense Satellite Communications System (DSCS), and Commercial Wideband Satellite Program (CWSP)).

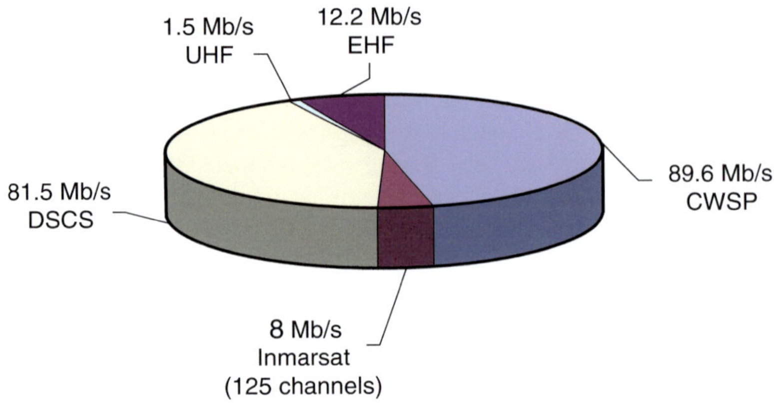

FIGURE 4.3 Fleet bandwidth (in megabits per second (Mb/s)) estimated for spring 2003. NOTE: A list of acronyms is provided in Appendix G. SOURCE: Data provided by the Program Executive Office for Command, Control, Communications, Computing, and Space at a presentation to the committee on August 26, 2003.

Protected EHF (Military Strategic, Tactical, and Relay Satellite (MILSTAR)) systems provided only about 12 Mb/s (or 6 percent) of the Navy's total capacity.

To understand how much bandwidth may be required by a large naval platform, the Naval Network Warfare Command (NETWARCOM) maintains a communications-requirements database, detailing current and future communications needs by data type, user, and so on. Figure 4.4 shows NETWARCOM's estimate, resulting from this analysis, of the bandwidth required by a Nimitz-class aircraft carrier (CVN) from 2003 through 2009. Although this analysis is of the wideband requirement, it would be only modestly augmented by addition of narrowband and secured communications capabilities.

Thus, the current requirement of about 8 Mb/s is anticipated to grow to about 25 Mb/s for a CVN over the next 6 years. This threefold increase is representative of the expectations of the operational warfighter's anticipated needs. The NETWARCOM analysis is based on a bottom-up, interview-based analysis of data types, linked to fleet needs and desires. According to NETWARCOM, the assessment was done prior to the requirements collection activities of FORCEnet and the collection of the associated needs of the three pillars of Sea Power 21.[21] The study also was likely influenced by the perception that available space-based bandwidth will be relatively fixed until 2009 when MUOS and the Advanced

[21]VADM Richard W. Mayo, USN, Commander, Naval Network Warfare Command, presentation to the committee, August 26, 2003.

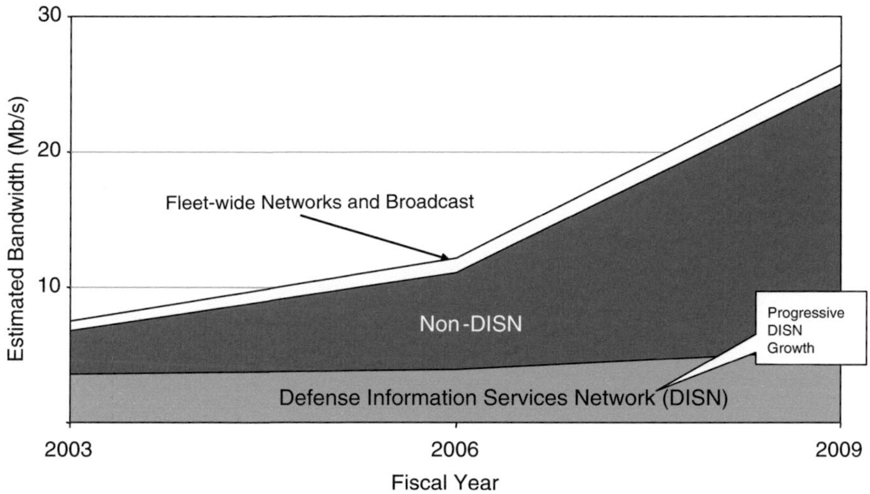

FIGURE 4.4 Future wideband communications needs for a Nimitz-class aircraft carrier. NOTE: Estimates assume duplex terminal operation. SOURCE: CAPT John Yurchak, USN, Naval Network Warfare Command. 2003. "C5I [command, control, communications, computers, combat systems, and intelligence] Day—Progress Report Fleet Satellite Network Communications," presentation to ADM Robert J. Natter, USN, Commander, Fleet Forces Command, Norfolk, Va., March 31.

Wideband System (AWS) are scheduled to enter operation. Finally, it appears to the committee that the Navy's input to the TCA is also based on this limited analysis.

Table 4.3 presents an estimate of the aggregate theater bandwidth that was available for naval forces in FY00, the actual theater bandwidth available to U.S. Central Command in FY03, and NETWARCOM's projection of future theater bandwidths available by FY07. The total available satellite bandwidth for a theater was approximately 22 Mb/s in FY00 and 49 Mb/s in FY03, with a potential of 318 Mb/s in FY07. Table 4.4 presents maximum bandwidth (in Mb/s) available for individual platforms (those that have been fitted to the current state of the art) in FY00 and FY03 as well as NETWARCOM's projection for FY07. Note that even in FY07, naval platforms will have rather low communications bandwidth. For example, a command ship (LCC) will have slightly more than 10 Mb/s, while an attack submarine (SSN) will have no more than 0.5 Mb/s of wideband capacity.

Note that the projected communications shortfalls will be particularly significant in cases in which command resides on a naval platform. As stated above, for example, during Operation Iraqi Freedom (spring 2003), the U.S. joint forces

TABLE 4.3 Total Theater Bandwidth Available for Naval Forces in FY00, Actual Theater Bandwidth Available to U.S. Central Command in FY03, and Naval Network Warfare Command's (NETWARCOM's) Projection of Future Theater Bandwidth Available by FY07

	System	Bandwidth (in Mb/s)		
		FY00 (Estimated)	FY03 (Actual)	FY07 (Projected)
Earth Beam	Defense Satellite Communications System (DSCS)	8.192	17.408	0.0
	Commercial Wideband Satellite-communications Program (CWSP)	10.752	30.208	0.0
	International Maritime Satellite (Inmarsat)	3.968	13.888	11.200
	Wideband Gapfiller System (WGS)			0.0
Spot Beam	Extremely High Frequency Medium Data Rate (EHF MDR)		0.0	14.080
	Global Broadcast Service (GBS)		49.152	49.152
	Wideband Gapfiller System (WGS)			244.480
TOTAL		22.912	110.656	318.912

SOURCE: CAPT John Yurchak, USN, Naval Network Warfare Command. 2003. "C5I [command, control, communications, computers, combat systems, and intelligence] Day—Progress Report Fleet Satellite Network Communications," presentation to ADM Robert J. Natter, USN, Commander, Fleet Forces Command, Norfolk, Va., March 31.

utilized over 750 Mb/s of communications on a sustained basis.[22] This level is approximately 50 times that being planned for any Navy command ship, even out to 2007. Since bandwidth utilization has historically grown exponentially (and there is no reason to believe that it has yet leveled off), this current gap will only grow by 2007. The committee believes that this problem will be particularly acute for the Sea Basing mission (as described above), since a joint forces command-and-control center may need to be based at sea and thus rely solely on space-based communications.

Communications Capability Gaps

In reviewing available information and in assessing the space-based communications capabilities required to implement the robust elements of Sea Power 21 described above, the committee noted several gaps and apparent issues that will

[22]Lt Gen T. Michael Moseley, USAF, Commander, Central Air Forces. 2003. *Operation Iraqi Freedom—By the Numbers*, Shaw Air Force Base, S.C., April 30, p. 12.

require concerted naval effort to resolve over the coming years. While the Navy is correct in projecting a general trend of bandwidth growth, the committee believes that the exponential growth in capability- and platform-generated data cause the current naval bandwidth projections to be severely underestimated. Further, the committee believes that the reliance of warfighting capability on satellite communications will necessitate new requirements to substantially increase assured and nonassured link availability.

The tactical and mobile user will require high-availability, high-bandwidth, assured communications links worldwide. To some extent this issue has been discussed and recognized by many in the Navy, but the Navy has not converted these issues into clear goals, requirements, or documentation. In particular, the requirements-based analysis of future needs, to date, has been largely derived by looking back at what has been the communications capability, with respect to data types, speeds, and sources. Because the analyses were developed prior to the clear articulation of the Sea Power 21 and TPPU concepts as well as before the experiences of Operation Iraqi Freedom, they could not be expected to support these new needs. Also, the current communications requirements have not accounted for the push of technological capability to provide new data types and capabilities (e.g., streaming video from multiple BLOS UAVs, the extreme bandwidth needs of the SBR, and the articulation of Sea Basing). Finally, the migration to TPPU from the previous concept of tasking, processing, exploiting, and

TABLE 4.4 Naval Platform Wideband Capacity Available for Naval Forces in FY00, Actual Naval Platform Wideband Capacity Available in FY03, and Naval Network Warfare Command's (NETWARCOM's) Projection of Naval Platform Wideband Capacity Available by FY07

	Wideband Capacity (Mb/s)		
Platform	FY00 (Estimated)	FY03 (Actual)	FY07 (Projected)
Command ship (LCC)	2.048	3.072	10.496
Aircraft carrier, nuclear-powered (CV/CVN)	2.048	3.072	8.448
Amphibious assault ship (LHD/LHA)	2.048	2.304	8.448
Dock landing ship/amphibious transport dock (LSD/LPD)	0.064	0.064	3.328
Guided cruiser (CG)	0.064	0.384	3.328
Guided missile destroyer (DDG)	0.064	0.128	3.328
Destroyer/guided missile frigate (DD/FFG)	0.064	0.064	3.328
Fast combat support ship (AE/AO/AF)	0.064	0.512	0.512
Attack submarine (SSN)	0.032	0.064	0.512
Guided missile attack submarine (SSGN)	NA	NA	0.768

NOTE: NA = wideband capability not available to platform. SOURCE: CAPT John Yurchak, USN, Naval Network Warfare Command. 2003. "C5I [command, control, communications, computers, combat systems, and intelligence] Day—Progress Report Fleet Satellite Network Communications," presentation to ADM Robert J. Natter, USN, Commander, Fleet Forces Command, Norfolk, Va., March 31.

disseminating (TPED) will require the reassessment of database locations, server sizes, and communications bandwidth across the GIG. The current integration of total space-based communications needs is being compiled by NETWARCOM and, it is hoped, will engage many of these issues.

Thus, it appears that the Sea Power 21 concept is inconsistent with current Navy plans and future requirements for space-related communications, and it appears that the Navy is not investing appropriately in thorough operational analysis to support these established requirements. Robust, high-bandwidth space-based communications will be essential for FORCEnet, but it is unclear whether planned improvements in naval space-based communications will satisfy even the minimal FORCEnet needs for bandwidth, availability, or information assurance. Table 4.5 summarizes some of these more important current gaps and resulting needs.

TABLE 4.5 Space-Based Communications Gaps and Resulting Needs for Navy Use

Space-Based Communications Needs	Gaps or Shortcomings
Gb/s bandwidth to platforms	Continual expansion of communications bandwidth is caused by: consolidation of Sea Power 21 pillars; technology push; transition to concept of tasking, posting, processing, and using; and new generations of high-data-rate sensors (SBR, UAV streaming video, and others).
Very high link availability	Current analysis shows nonavailability of instantaneous tactical-user communications connectivity ranging (minute by minute) from 5 to 30 percent of the time. As ISR cueing migrates to beyond-line-of-sight (BLOS) ranges, this will be unacceptably low.
Very high communications assurance	Most space-based communications bandwidth is now supplied by unprotected DOD and commercial systems. Currently there is no parsing of requirements into assured versus unprotected bandwidth needs and no analysis of how to perform operations in the event of losing access to unprotected communications.
Seamless integration with the Global Information Grid	Naval communications are largely channelized, including architectures for MUOS and WGS. The Transformational Communications Architecture will strongly migrate to IP-based architecture, and will require a large shift in communications philosophy, systems, operations, training, and platform systems.
Very low latency worldwide communications	Beyond-line-of-sight cueing for theater and strategic missile warning and defensive systems will require revision in many communications concepts, infrastructure, and capabilities.

NOTE: A list of acronyms is provided in Appendix G.

While the Navy is supporting S&T development in some of the areas indicated in Table 4.5, it would be unwise to focus all naval space S&T into a space-based communications research program. Virtually all aspects of space-based communications will be far more useful if combined into hybrid systems (e.g., combined space and airborne networks), so it is important also to devise S&T activities that cross the space/nonspace boundary. Finally, it is noted that many productive research programs, including several current ones, do not issue from a top-down analysis of needs, but from the understanding of technical requirements by capable technologists who propose innovative and unconventional ways to apply technology or provide new approaches to difficult problems; such efforts at innovation and technology push are encouraged.

Findings and Recommendations Regarding Space-Based Communications

Basic Communications Capabilities

For the naval forces of the future to be effectively engaged in large-scale planning, command and control, and ISR operations as part of the joint forces, naval capability requirements for future fleet space-based communications appear to be significantly underestimated in the following areas:

- Total fleet bandwidth requirements,
- Individual platform bandwidth requirements, and
- Availability and assurance of GIG communications for mobile/tactical users.

The shortcomings of official requirements estimates are recognized by the Navy and are likely being accounted for in FORCEnet requirements studies now under way.[23] However, a large mismatch currently exists between the FORCEnet and Sea Basing needs and the current bandwidth planning estimates. In addition, the evolving concept of TPPU may drive increases in communications bandwidth, processing power, and interpretation functionality at the user end of the GIG.

Much of the Navy's current operational communications capability is dependent on nonprotected space-based communications systems; hence, loss or adversary exploitation of this bandwidth could have significant consequences during naval operations. While the Navy relies on MILSTAR and its successor, the Advanced Extremely High Frequency (AEHF) satellite, for assured communications, it is unclear if current and planned assured communications bandwidths

[23]RADM Jay Cohen, USN, Chief of Naval Research, presentation to the committee, July 29, 2003.

that could be made available to the Navy are sufficient to sustain core naval operations in the event that unprotected communications assets were disrupted.

Thus, current naval platforms suffer from fragile and intermittent network connectivity. Connectivity and communications availability are currently limited by antenna multiplicity, placement, and real estate issues related to the design of the platform superstructures. Implementing the Sea Power 21 concepts will require near-continuous communications between commanders and tactical forces, and the current level of communications outages due to antenna geometry and placement will become increasingly unacceptable.

Recommendation 4.10. The Department of the Navy should increase its depth of understanding of Navy and integrated joint future communications needs.

The Department of the Navy cannot remain relatively passive in accepting the Department of Defense's (DOD's) space-based communications systems capabilities. The Navy should conduct its own systems analysis of communications requirements for a 20-year period on the basis of Sea Power 21 concepts. This function should be ongoing; it should be done in conjunction with a set of developed and maintained mission scenarios over the moving window of planning periods as technology, warfighting concepts, and threat understanding evolve. In particular, such an analysis program should include the following:

- Input from the space-based-communications, information-assurance, and science and technology communities, as well as from warfighters, to help ensure that potential limitations and future capabilities are included in the analysis;
- Regular interaction with large-scale experimental testbeds, including the evolving Transformational Communications testbed being developed by the Naval Research Laboratory;
- Investigation of the partitioning of requirements between space- and ground-based systems and additionally among various space-based systems, independent of current program management—the investigation should also be revalidated periodically to ensure that it is current with warfighting, weapons, sensors, and threat analyses and should serve as a basis for participation in the development of detailed requirements for all space-based communications acquisition programs undertaken by the Navy, Air Force, or other DOD agencies;
- Review of future terminal and antenna configurations and strategies in order to develop a long-range strategy to consolidate antennas, terminals, and network interface electronics into an efficient, continuous interface to the Global Information Grid for naval platforms;
- Regular red teaming to ensure that space-based communications requirements are consistent with warfighting strategy, new systems concepts, and evolving technology; and
- A comprehensive account of naval and joint warfighting operations and of technology evolution. Specific elements should include the following:

—Joint operations having extensive involvement with the DOD Executive Agent for Space as well as with other Services;

—Sea Strike, Sea Shield, Sea Basing, and FORCEnet capability needs;

—Increasingly large data volumes expected to be produced by emerging airborne and space-based sensor capabilities, including Space Based Radar, the Future Imagery Architecture, and unmanned aerial vehicle constellations;

—Partitioning of communications requirements by assurance levels to enable a warfighting core capability should unprotected communications prove vulnerable; and

—Advanced data and technology concepts, including the needs associated with the tasking, posting, processing, and using concept.

Research and Development Programming

Although the Navy has several proposed communications projects targeting the bandwidth gaps discussed above, the current R&D program is underfunded and appears unlikely to produce the technology base necessary to enable the acquisition of systems that will reliably and continuously connect tactical naval users to the GIG. Naval communications requirements derived from Sea Power 21 concepts rely on such global connectivity. The Chief of Naval Research (CNR) recently stated that "extremely high data rates using laser communications will not be available to the Navy tactical user without technology development for the 'final mile' to the fleet."[24] Thus, reliable, extremely wide bandwidth communications connectivity is the most significant projected Sea Power 21 capability gap. The Navy currently lacks the space-based communications operational analysis needed to assess future technical options and their maturity. Without such an analytic basis, the Navy will find it increasingly difficult to interact with and influence DOD programs (such as TCA) and judge the degree of reliance that the Navy will need to place on such programs during its own strategic and operational planning.

Recommendation 4.11. The Department of the Navy should fund and manage an expanded operational analysis program focused on supporting research and development in space-based communications.

The expanded program should focus on developing solutions in order to accomplish the following: (1) provide multiple gigabit-per-second-class bandwidth, connecting mobile Navy users to the Global Information Grid; (2) provide high availability for all user platforms; (3) provide high assurance for all user platforms; (4) resolve antenna and terminal multiplexing issues; and (5) ensure that Navy-led space-based communications programs such as the Mobile User

[24]RADM Jay Cohen, USN, Chief of Naval Research, presentation to the committee, July 29, 2003.

Objective System, as well as shipboard and ground-based networks, evolve to be fully compatible with the GIG and the Distributed Common Ground System transport standards. The Department of the Navy should also allocate funding for basic communications R&D to ensure that new technologies and concepts are available in the future.

Recommendation 4.12. The Department of the Navy not only should support research and development programs, but also should support experimental programs aimed at supporting space-based communications.

In particular, the Department of the Navy should consider supporting a space experiment (perhaps through the Advanced Concept Technology Demonstration program) to demonstrate high-availability communications at gigabit-per-second rates from space to a deployed naval platform. Such an effort should be considered together with current and proposed Defense Advanced Research Projects Agency and Air Force programs in optical communications from space, with a suggested naval research role in supplying the last-mile link from space systems to the fleet or other mobile naval tactical users. This experimental program and analysis should include verification of the end-to-end Transformational Communications (TC) Global Information Grid Bandwidth Expansion concept using the Naval Research Laboratory's TC testbed. This testbed is already being organized by the Naval Research Laboratory for use with TC as well as for use by other defense agencies. It would be wise for naval communications studies and architecture development teams to take advantage of the testbed's existence.

Recommendation 4.13. The Department of the Navy should direct research and development aimed at the problem of low-latency communications from space-based sensors to platforms, particularly with respect to the cueing of fast-moving targets from beyond-line-of-sight sensors and national systems. Such an activity should be done in conjunction with improvements to the Cooperative Engagement Capability as well as other missile defense efforts.

Recommendation 4.14. The Department of the Navy should focus more science and technology efforts on consolidated antenna and terminal configurations necessary to enable near-100-percent-reliable shipboard communications.

Recommendation 4.15. The Department of the Navy should support a naval space-based communications challenge and fund its science and technology (S&T) community to aggressively anticipate potential future space-based communications requirements.

For example, a suitable challenge to the S&T community is to demonstrate from space worldwide, 40 Gb/s connectivity to naval platforms with near-100-percent-available, high-assurance communications connectivity.

Narrowband Communications

Historically the Navy has been a leader in supplying national capability for communications to the tactical warfighter. It is continuing this leadership in the Mobile User Objective System program. By leading, staffing, and executing the study, specification, acquisition, activation, and operation of narrowband space-based communications systems, the Navy has an opportunity to invest the naval space cadre with on-the-job training and experience in space technology, issues, contracting, and operations. This experienced core, in turn, will increase the Navy's effectiveness as a partner in other National Security Space systems acquisitions, while ensuring that the acquired narrowband capability meets the Navy user's needs over the long term.

Recommendation 4.16. The Department of the Navy should continue its role as lead agency for narrowband communications. The Department of the Navy should direct the Mobile User Objective System (MUOS) program to direct special attention in FY05 to ensuring that MUOS will interface effectively as an edge system in the Transformational Communications Architecture, and to harden the system, as is feasible within cost and schedule constraints, against the evolving counterspace threat environment.

Recommendation 4.17. The Department of the Navy should revise its strategy of relying largely on commercial and unprotected communications during conflict. The Navy should carefully review the nature of potential threats to unprotected communications, both ground- and space-based, and take these threats into account when specifying next-generation communications needs and requirements. The Navy should also determine its core warfighting communications capability needs and should specify robust protection for these minimum capabilities to ensure adequate communications capabilities in the event of a total loss of access to commercial systems.

Navy Participation in National Security Space Activities

The leadership of the Navy in defining the Transformational Communications Architecture is established, but its increased participation in evolving the system concept is essential in providing the technology base and the system definition, development, and acquisition.

Recommendation 4.18. The Department of the Navy should increase its personnel assignments to support the Transformational Communications Architecture program. The Department of the Navy should allocate naval personnel so that on the order of 10 to 15 percent of the total military and support staffing of this major acquisition program is represented by the Naval Services.

POSITION, NAVIGATION, AND TIMING

Effective military operations extending across the entire spectrum of warfare require a robust and accurate system for position, navigation, and timing (PNT). Space-based navigational systems (GPS in particular) use satellites to allow users to establish three-dimensional positions (latitude, longitude, and altitude) of airborne and terrestrial platforms and to coordinate precision time and time-interval measurements. Such navigational systems have become the predominant means for providing vital military information. This fairly recent move away from long-wave radio navigation and timing systems (such as the Long Range Navigation (LORAN) system) and celestial observations has influenced military operations well beyond original expectations. Highly accurate clocks and frequency sources are now of vital importance to the DOD, because the accuracy and stability of these devices are key determinants of the performance of command, control, communications, and intelligence; navigation; surveillance; electronic warfare; missile guidance; identification-friend-or-foe systems; and precision military operations.

Background

Transit

The first satellite navigation system, the Navy's navigation satellite system—Transit—had its inception just days after the former Soviet Union launched Sputnik on October 4, 1957.[25] The idea for Transit came about when scientists at the Applied Physics Laboratory (APL) at Johns Hopkins University were able to determine Sputnik's orbit by analyzing Doppler shifts of its radio signals measured during a single pass. Frank McClure, then-chairman of APL's research center, later suggested that if the satellite's position were known and predictable, then the measured Doppler shift could be used to locate a receiver on Earth—in other words, one could navigate by satellite. Under sponsorship by the Navy's Strategic Programs Office and the Advanced Research Projects Agency (ARPA), APL began developing the Transit system in 1958; the system became operational in 1964.[26]

Transit was originally developed to provide accurate, reliable, all-weather, global navigation for use by ballistic-missile-carrying submarines. Transit's use spread to surface vessels, and in 1967 the system was released for public and

[25]Additional detail on the Navy's development of space is provided in Appendix A.

[26]Johns Hopkins University, Applied Physics Laboratory. 1996. *The Legacy of Transit*, Laurel, Md.; and National Research Council. 2002. *An Assessment of Precision Time and Time Interval Science and Technology*, The National Academies Press, Washington, D.C.

commercial use by ships of all friendly nations. Approximately 28 Transit-series satellites were launched during the lifetime of the program, and an 8-satellite constellation was still operating when the DOD phased out its use as a navigational system on December 31, 1996.

Timation

In 1964, NRL put forth a new concept for an improved space-based navigation system. This system would involve time (or range) measurements between a satellite and a user that were based on the utilization of spaceflight-qualified precision clocks. It was predicted that timing signals from such a satellite could provide more precise navigation than was available from Transit, as well as supplying a uniform global time standard. To achieve this goal, NRL started programs to develop improved quartz frequency standards suitable for spaceflight. Soon thereafter, the Timation program, which relied on atomic clocks in space, was established. Three satellites were launched during the experimental Timation program. The third Timation satellite was renamed Navigation Technology Satellite (NTS) 1 and flew the first atomic clock in 1974. Later, in 1977, NTS-2 was launched and flew the first cesium clock in space. These space-qualified atomic clocks were then used in the next-generation satellite navigation system, the Navigation Satellite Timing and Ranging/Global Positioning System (NAVSTAR/GPS), more commonly known as GPS.[27]

NAVSTAR Global Positioning System

The position, navigation, and timing system known today as NAVSTAR/GPS (or just GPS) capitalized on the several satellite navigation systems and concepts developed by or for the DOD, including Transit and Timation, and additional Air Force (Project 621B) and other DOD-wide studies. By 1972, the best characteristics of each of these programs had coalesced to form the general system characteristics and initial design parameters for GPS. From its inception, GPS was designed to meet the radio navigation requirements of all of the military Services as well as those of civilian users. On February 22, 1978, the Air Force began launching experimental GPS satellites, termed Block I satellites. After the third satellite successfully achieved orbit, testing of the system began. Using a portable receiver mounted in a truck moving 80 kilometers per hour, the Air Force showed that the desired positioning accuracy of 10 meters in two dimen-

---

[27]Gary Federici, Robert Hess, and Kent Pelot. 1997. *From the Sea to the Stars: A History of U.S. Navy Space and Space-Related Activities,* Working Paper, The Center for Naval Analyses, Alexandria, Va.

sions was easily achievable. After tests with the first three experimental satellites proved successful, eight additional Block I satellites were launched to complete the design and testing phase of the GPS program. Although these satellites were intended to have a 3-year life span, they achieved an average operational life of almost 7 years.[28]

GPS relies on the principle of pseudo-ranging to provide accurate positioning to its users. Each satellite in orbit continuously transmits a radio signal with a unique code, called a pseudo-random noise (PRN) code, that includes data about the satellite's position and the exact time that the coded transmission was initiated, as kept by the satellite's onboard atomic clock. (GPS utilizes Coordinated Universal Time maintained by the U.S. Naval Observatory, in Washington, D.C.) A pseudo-range measurement is created by measuring the distance between a user's receiver and a satellite by subtracting the time at which the signal was sent by the satellite from the time at which it is received by the user. Once three ranges (or distances) from three known positions are measured, a position in all three dimensions can be determined. In the case of GPS, however, a fourth satellite is generally needed in order to eliminate a common bias in the pseudo-ranges to all satellites caused by a lack of synchronization between the satellite and receiver clocks. Once this clock bias is eliminated by the presence of a fourth signal, a highly accurate three-dimensional position can be determined.

Instead of transmitting one PRN code on one radio signal, each GPS satellite actually transmits two distinct spread spectrum signals that contain two different PRN codes, called the coarse acquisition (C/A) code and the precision (P) code. The C/A-code is broadcast on the L-band carrier signal known as L_1, which is centered at 1572.42 MHz. The P-code is broadcast on the L_1 carrier in phase quadrature with the C/A carrier and on a second carrier frequency, designated as L_2, that is centered at 1227.60 MHz.

The L_1 C/A-code provides free positioning and timing information to civilian users all over the world; it is known as the standard positioning service (SPS). The timing information on the C/A-code is also used by some receivers to aid in the acquisition of the more accurate P-code. The P-code is normally encrypted using National Security Agency (NSA) cryptographic techniques, and decryption capability is available only to the military and to other authorized users. When encrypted, the P-code is normally referred to as the Y-code. Y-code availability through authorized decryption capability is known as the precise positioning service (PPS).[29]

---

[28]National Research Council. 1995. *The Global Positioning System: A Shared National Asset*, National Academy Press, Washington, D.C.

[29]National Research Council. 1995. *The Global Positioning System: A Shared National Asset*, National Academy Press, Washington, D.C.

Before the PPS and SPS were officially established, GPS designers had anticipated that use of the Y- and C/A-codes would produce very different levels of positioning accuracy. Use of the Y-code was expected to result in 10 m accuracy, whereas the C/A-code was expected to provide accuracy of 100 m. Developmental testing of the Block I GPS satellites, however, showed that the accuracy difference between the two codes was not this significant. PPS accuracy was officially specified as 16 m spherical error probable (SEP), and SPS accuracy was set at 100 m SEP. This two-level accuracy arrangement was made possible on the Block II/IIA satellites through an accuracy denial method known as selective availability (SA). SA is a purposeful degradation in GPS navigation and timing accuracy that controls access to the system's full capabilities. SA was accomplished in part by intentionally varying the precise time of the clocks onboard the satellites, which introduces errors into the GPS signal. An extensive study conducted by the National Research Council (NRC) in 1995 determined that the military effectiveness of SA was significantly undermined by differential GPS augmentations.[30] For this and other, related civil-use reasons, the NRC recommended that SA be turned to zero, but the capability is retained for potential emergency use by the National Command Authority. By Executive Order in 2000,[31] SA was turned to zero on the C/A-code L_1 signal. SA and a new anti-spoofing encryption are still retained for the military Y-code.

Differential GPS (DGPS) is the most widely used method of GPS augmentation; it can significantly improve the accuracy and availability of basic GPS. DGPS is based on knowledge of the highly accurate, geodetically surveyed location of a GPS reference station, which observes GPS signals in real time and compares their ranging information to the ranges expected to be observed at the DGPS fixed reference location. The differences between observed ranges and predicted ranges are used to compute corrections to GPS parameters, error sources, and/or resultant positions. These differential corrections are then transmitted to GPS users, who apply the corrections to their received GPS signals or computed position. As an example, the Wide-Area Augmentation System is a wide-area DGPS concept planned by the Federal Aviation Administration to improve the accuracy, integrity, and availability of GPS to levels that support flight operations in the National Airspace System, from en route navigation through Category I precision approaches.

---

[30]National Research Council. 1995. *The Global Positioning System: A Shared National Asset,* National Academy Press, Washington, D.C.

[31]*Statement by the President Regarding the United States' Decision to Stop Degrading Global Positioning System Accuracy*, May 1, 2000. Available online at <http://www.ostp.gov/html/0053_2.html>. Accessed February 19, 2004.

GPS user equipment varies widely in cost and complexity, depending on the receiver design and application. Receiver sets, which currently vary in price from less than $400 to over $30,000, can range from simple, one-channel devices that only track one satellite at a time and provide only basic positioning information, to complex, multichannel units that track all satellites in view and perform a variety of functions. Currently, GPS is the sole U.S. navigation satellite system for civil and military use.

The Global Orbiting Navigation Satellite System

The Russian Global Orbiting Navigation Satellite System (GLONASS) was developed in the early to mid-1980s by the military of the former Soviet Union. The GLONASS space segment also is designed to consist of 24 satellites, but these satellites are to be arranged in three orbital planes at 19,100 km (11,870 mi) altitude and at an inclination of 64.8 degrees rather than in six planes as for GPS. The full GLONASS constellation was scheduled to be operational in 1995.[32] Currently, 11 GLONASS satellites are on-orbit.

GLONASS differs from GPS in the way that the user segment differentiates one satellite from another. Instead of each satellite's transmitting a unique PRN code as GPS satellites do, GLONASS satellites all transmit the same PRN code on different channels or frequencies. All of these frequencies, however, are in the L band near either of the two GPS downlink signals, which simplifies the task of designing integrated receivers. In addition, GPS and GLONASS use different time standards and coordinate systems. Discrepancies between these standards and coordinates exist and must be corrected by combined receivers if an integrated GPS/GLONASS capability is desired.

Galileo

In May 2003, the European Space Agency initiated a program to develop a global satellite navigation system (Galileo) with capabilities similar to those of GPS. However, the European Space Agency has asserted that Galileo will be built for civilian purposes, with sufficient positional and timing accuracy to support Europe's integrated transport system. Galileo will consist of about 30 satellites (27 operational and 3 spares), positioned in three circular orbital planes at 23,616 km altitude and at an inclination of 56 degrees. Galileo will provide dual downlink frequencies in the L band, with a high degree of availability through an integrated failure-reporting system. The initial Galileo validation launches are planned for 2005 through 2006. Once the on-orbit validation phase has been

---

[32]National Research Council. 1995. *The Global Positioning System: A Shared National Asset*, National Academy Press, Washington, D.C.

completed, the remaining satellites will be launched to reach a full operational capability in 2008.

Galileo will provide an additional search-and-rescue feature: each satellite will be equipped with a transponder that will be able to transfer distress signals from users' transmitters to a Rescue Coordination Center for the activation of a rescue operation. At the same time, the system will provide a signal to the user, indicating that the situation has been detected and that help is being provided.

It is intended that the Galileo user's receiver equipment be interoperable with GPS and GLONASS.

Current and Planned Capabilities

The NAVSTAR GPS is recognized as the only satellite navigation system currently employed by the U.S. military and civil sectors. GPS is viewed as a key enabler to DOD transformation, as its precision positioning and timing capabilities enable continuous situational awareness, precision strike, autonomous operations, and precision synchronization of combat operations.[33]

Space Segment

A space-based radio positioning system nominally consisting of a 24-satellite constellation, GPS provides navigation and timing information to military and civilian users worldwide. The constellation currently consists of Block II, IIA, and IIR satellites, arranged in six orbital planes of 55-degree inclination, at 10,988 nautical miles altitude. Each satellite completes one orbit in one-half of a sidereal day and therefore passes over the same location on Earth once every sidereal day (approximately every 23 hours and 56 minutes). The particular orbital configuration and number of satellites allows a user at any location on Earth to have at least four satellites in view 24 hours per day. Each Block II/IIA satellite is designed to operate for 7.5 years; the Block IIR satellites' design life is 10 years. Block II/IIA provides SPS with a 16 to 24 m SEP on the C/A code L_1 signal and PPS with a 16 m SEP on the P-code L_1 and L_2 signals.[34]

Block IIRs began replacing Block II/IIAs on July 22, 1997. There are currently eight Block IIR satellites on orbit, with the next launch planned for October 2003.[35] Block IIR satellites boast dramatic improvements over the previous blocks. They also have reprogrammable satellite processors enabling in-flight

[33]Global Positioning System Joint Program Office (GPS JPO). 2003. *NAVSTAR GPS—Providing DOD Transformation*, Washington, D.C., October 28.

[34]Global Positioning System Joint Program Office (GPS JPO). 2003. *NAVSTAR GPS—Providing DOD Transformation*, Washington, D.C., October 28.

[35]Global Positioning System Joint Program Office (GPS JPO). 2003. *NAVSTAR GPS Overview*, March.

problem fixes. Eight Block IIR satellites are being modified to radiate the new, military-only (M-code) signal on both the L_1 and L_2 channels, as well as the more robust civil signal (L_2C) on the L_2 channel. The M-code signal is a more robust and capable signal architecture. It will have increased power and reduced vulnerability to signal jamming. In addition to the improved signals, the reliability of the GPS navigation message will be improved by adding more satellite monitoring stations. These additional stations will ensure that at least two stations will be able to simultaneously monitor each satellite. The data collected by these additional stations will then be combined with the data from the existing monitoring stations and sent to the master control station for processing. The result is improved accuracy of the navigation message broadcast by the satellite. The Block IIR positional accuracy today is 6.6 m. The first modified Block IIR (designated as the IIR-M) is planned for launch in 2004.[36]

Block IIF satellites are the next generation of GPS space vehicles. Block IIF provides all of the capabilities of the Block IIR-M with some additional benefits as well. Improvements include an extended design life to 12 years, faster processors with more memory, and a new civil signal on a third frequency (L_5). Block IIF positional accuracy is planned to be 3.4 m. The first Block IIF satellite is scheduled for launch in 2006.[37]

Control Segment

The GPS operational control segment (OCS) consists of the master control station, located at Schriever Air Force Base, Colorado; remote monitoring stations, located in Hawaii, Diego Garcia, Ascension Island, and Kwajalein; and uplink antennas located at three of the four remote monitoring stations and at the master control station. The four remote monitoring stations contribute to satellite control by tracking each GPS satellite in orbit, monitoring its navigation signal, and relaying this information to the master control station. The master control station is responsible for overall satellite command and control, which includes maintaining the exact orbits of each satellite and determining any timing errors that may be present in the highly accurate atomic clocks aboard each satellite.

Major improvements are planned for the OCS. They include a new master control station with improved operator interfaces and Block IIR/IIF capabilities at Schriever Air Force Base, an alternate master control station at Vandenberg Air Force Base, and the establishment of additional monitoring sites at NGA locations around the globe.[38]

---

[36]Global Positioning System Joint Program Office (GPS JPO). 2003. *NAVSTAR GPS—Providing DOD Transformation*, Washington, D.C., October 28.

[37]Global Positioning System Joint Program Office (GPS JPO). 2003. *NAVSTAR GPS Overview*, Washington, D.C., March.

[38]Global Positioning System Joint Program Office (GPS JPO). 2003. *NAVSTAR GPS—Providing DOD Transformation*, Washington, D.C., October 28.

Modernized User Equipment

Another effort under way at the GPS Joint Program Office (JPO) is that of developing a single modernized GPS receiver card to demonstrate M-code receiving and processing capability. The Services are expected to procure their own modernized user equipment based on the demonstrated performance of the M-code receiver card and to integrate this capability into their respective specific platform host receivers. The M-code card development effort is planned for completion by February 2008. To assist the GPS JPO with these activities, the DOD Executive Agent for Space recently requested that each of the Service departments conduct a GPS user equipment synchronization study to ensure that M-code user equipment development is synchronized with space- and ground-segment M-code capabilities.

Global Positioning System Block III

A future positioning system, the Block III Global Positioning System (GPS III), is in the planning stages. This system is intended to provide for the assured delivery of enhanced PNT signals and to offer related services to meet the needs of the next generation of GPS users. The GPS III program includes an integrated space-segment and control-segment system that incorporates the nuclear detonation detection system, a security infrastructure to provide user access to and protection of the entire system, and the incorporation of additional mission capabilities (including blue force tracking, search and rescue, and others).

GPS III is envisioned to involve a total reexamination of the GPS architecture in order to achieve increased performance in hostile electromagnetic environments and possibly to feature the use of satellite-to-satellite links to improve control efficiencies. GPS III is also planned to incorporate a third civil signal and the use of spot beams to enable higher effective radiated power in small warfare theaters. It has been established that GPS III will have a minimum of +20 dB gain over Block IIA/IIR capabilities, with a long-term objective of +27 dB to achieve significant performance in jamming environments.[39] GPS III positional accuracy is expected to be 1.9 m.

The statement of objectives for GPS III was released in August 2001, as part of a system definition and risk reduction announcement by the GPS JPO.[40] The first launch of a GPS III satellite is planned for 2012, with a full operational capability set for 2018.

---

[39]Global Positioning System Joint Program Office (GPS JPO). 2003. *NAVSTAR GPS—Providing DOD Transformation*, Washington, D.C., October, 28.

[40]Global Positioning System Joint Program Office (GPS GPO). 2001. *Global Positioning System (GPS) III, System Definition and Risk Reduction (SD/RR), Statement of Objectives (SOO)*, Washington, D.C., August 30.

Role of the Department of the Navy

Assigned the primary lead in PNT for the Navy, the Oceanographer of the Navy is responsible for providing resource and program sponsorship. In addition, the Fleet Forces Command is responsible for forwarding the fleet's GPS and PNT requirements, the Space and Naval Warfare Systems Command (SPAWAR) is assigned to develop and acquire naval GPS user equipment, ONR supports R&D on precision clocks, NRL monitors all GPS clock performance, and the U.S. Naval Observatory (USNO) maintains the master time reference for the DOD.

Requirements

The Navy, together with the other Services, participates in developing the functional and performance requirements for the GPS constellation, based on a number of military applications. Table 4.6 provides a summary of naval applications and associated positioning and radio frequency interference (RFI) resistance requirements. Table 4.7 provides a summary of military aviation and precision-guided munitions applications and associated positioning and RFI resistance requirements. Both of these tables were published in 1995 in the National Research Council report *The Global Positioning System: A Shared National Asset*,[41] and although somewhat dated, they provide a good indication of requirements placed on GPS for positional accuracy, integrity, and RFI resistance. In addition, Appendix C of this report provides an evaluation of Sea Power 21 dependencies on current and projected PNT capabilities. The overall result of this comparison shows that precision PNT is critical for most Sea Power 21 military operations identified.

User Equipment

SPAWAR and the GPS and Navigation Systems Division of the SPAWAR Systems Center-San Diego, are the Navy's principal acquisition and engineering development activities for GPS receiver development and integration. These organizations have provided technical, management, and engineering support for GPS receivers deployed on more than 120 Navy, Marine Corps, and Coast Guard platforms. SPAWAR Systems Center-San Diego also manages the GPS Central Engineering Activity Laboratory for evaluating GPS receivers.

Timing Standards

NRL, ONR, and USNO continue to maintain their leadership role in the research and development of precision time standards. As noted earlier, through

[41] National Research Council. 1995. *The Global Positioning System: A Shared National Asset*, National Academy Press, Washington, D.C.

TABLE 4.6 Naval Applications and Associated Positioning and Radio Frequency Interference (RFI) Resistance Requirements

	Application	Accuracy	RFI Resistance
En route	Pilotage and coastal waters	72.0 m CEP	High
Navigation	Inland waters	25.0 m CEP	High
	Open waters	2400.0 m CEP	High
	Rendezvous	380.0 m CEP	High
	Harbor	8.0 m CEP	High
Mine warfare	Swept channel navigation and defensive mining	16.0 m CEP	High
	Offensive mining	50.0 m CEP	High
	Antimine countermeasures	<5.0 m CEP	High
	Geodetic reference guide	128.0 m CEP	High
Special warfare	Airdrop	20.0 m CEP	High
	Small craft	50.0 m CEP	High
	Combat swimming	1.0 m CEP	High
	Land warfare and insert/extraction	1.0 m CEP	High
Task group operations	General task group operations	72.0 m CEP	High
Amphibious warfare	Beach surveys	185.0 m CEP	High
	Landing craft	50.0 m CEP	High
	Artillery and reconnaissance	<6.0 m CEP	High
Surveying	Hydrographic	<5.0 m (2 drms)	High
	Ocean and geophysical deep ocean	90.0 m (2 drms)	High
	Oceanographic	100.0 m (2 drms)	High

NOTE: CEP, or circular error probable, represents an accuracy that is achievable 50 percent of the time in two dimensions (latitude and longitude). Drms, or distance root mean square; 2 drms = 2.4 × CEP. SOURCE: National Research Council. 1995. *The Global Positioning System: A Shared National Asset*, National Academy Press, Washington, D.C.

its Timation program NRL made significant contributions to the development of precision frequency standards suitable for spaceflight. NRL became a key participant in the development of advanced atomic clocks for flight in GPS satellites.[42] Navy responsibility in precision time is currently designated by DOD Instruction 5000.2, Part 7, Section C, which calls for the Department of the Navy to carry out the following:

- Maintain the DOD reference standard through the USNO.
- Serve as the DOD precise time and time interval (frequency) manager, with responsibilities for

[42]National Research Council. 2002. *An Assessment of Precision Time and Time Interval Science and Technology*, The National Academies Press, Washington, D.C.

TABLE 4.7 Military Aviation and Precision-Guided Munitions Applications and Associated Positioning and Radio Frequency Interference (RFI) Resistance Requirements

	Application	Accuracy	RFI Resistance
Aviation	Low-level navigation and air drop	50.0 m (2 drms)	High
	Non-precision sea approach/landings	12.0 m (2 drms)	High
	Precision approach/landings (unprepared surface)	12.5 m (2 drms)	High
	Precision sea approach/landings	0.6 m (2 drms)	High
	Amphibious and antisubmarine warfare	50.0 m CEP	High
	Anti-air warfare	18.1 m CEP	High
	Conventional bombing	37.5 m CEP	High
	Nuclear bombing	75.0 m CEP	High
	Close air support/interdiction	9.0 m CEP	High
	Electronic warfare	22.5 m CEP	High
	Command, control and communications	37.5 m CEP	High
	Air refueling	370.0 m CEP	High
	Mine warfare	16.0 m CEP	High
	Reconnaissance	18.1 m CEP	High
	Magnetic and gravity survey	20.0 m CEP	High
	Search and rescue/medical evacuation	125.0 m CEP	High
	Mapping	50.0 m CEP	High
Precision-guided munitions		3.0 m CEP	High

NOTE: CEP, or circular error probable, represents an accuracy that is achievable 50 percent of the time in two dimensions (latitude and longitude). Drms, or distance root mean square; 2 drms = 2.4 × CEP. SOURCE: National Research Council. 1995. *The Global Positioning System: A Shared National Asset*, National Academy Press, Washington, D.C.

—Developing an annual DOD-wide summary of precise time requirements, and

—Coordinating the development of precise time and time interval techniques among DOD components.[43]

In addition to maintaining the DOD's master clock, the USNO has an active research effort in clock development, timescale algorithms, and time transfer. NRL also maintains Navy expertise in space clock technology, providing services and advice to Navy and DOD programs related to the space-based clocks used in GPS and other systems.[44]

---

[43]Department of Defense. 2002. DOD Instruction 5000.2, "Operation of the Defense Acquisition System," Part 7, Section C, E. Andrews, Jr., J. Stenbit, and T. Christie, April 5.

[44]National Research Council. 2002. *An Assessment of Precision Time and Time Interval Science and Technology*, The National Academies Press, Washington, D.C.

Findings and Recommendations Regarding Position, Navigation, and Timing

Precision position, navigation, and timing are critical to the Navy's new operational concepts embodied in Sea Power 21. An important factor is the global nature of these concepts. Naval forces arrayed over thousands (even tens of thousands) of miles rely heavily on space-derived PNT to coordinate and execute these operations effectively. The U.S. Navy has filled an important role in the research and development of much of the technology that led to today's highly capable satellite navigation systems and is currently engaged in the development of new technology for precision, space-qualified time standards.

NAVSTAR GPS continues to be the premier spaceborne capability for military and civilian users of precision PNT. An ambitious program is under way to replenish and upgrade this system with advanced-technology spacecraft and a modernized ground control system to better serve the needs of military and civil users. The new GPS M-code signal will have a significant impact on the accuracy and reliability of the system. The GPS Joint Program Office has an important technology development effort under way to demonstrate a single GPS receiver card for M-code receiving and processing. All military Services are effectively involved in and working with the GPS JPO in the evolution of future-generation GPS III capabilities. Each Service also conducts receiver development programs that are keyed to GPS space- and ground-segment developments.

Recommendation 4.19. The Department of the Navy should retain close ties with the Global Positioning System (GPS) Joint Program Office during the development of upgraded GPS space and ground segments. The Department of the Navy should also ensure that specific applications of integrated GPS (precision weapons systems, for example) are coupled to spacecraft capabilities that affect the resistance of these systems to radio-frequency interference (jamming). The Department of the Navy should conduct trade-off studies to determine the most cost-effective approach and strategy in developing guidance systems that rely on a combination of GPS and inertial guidance capabilities.

Recommendation 4.20. The Department of the Navy should initiate a GPS synchronization study similar to that being conducted by the Air Force to ensure that M-code (military-only) user equipment development is synchronized with space- and ground-segment M-code capabilities.

Recommendation 4.21. The Department of the Navy should sustain support to continue research and development in the area of precision timing standards and time transfer techniques, especially for potential use in future GPS space systems.

SPACE CONTROL

As increased reliance, if not outright dependence, on space capabilities becomes more widespread, space superiority and/or space control capabilities become more critical. While many of the details regarding space control have been and will remain classified, it is clear that naval forces will need the same level of security for space support as is needed by other joint forces. It is also clear that the Navy is well positioned to support the space control mission.

The space control mission area includes the elements of space situational awareness, defensive counter-space, offensive counter-space, and related battle management and command and control. Space situational awareness provides predictive battlespace awareness. Defensive counter-space provides protection for friendly space capability. Offensive counter-space provides the ability to disrupt, degrade, deny, or destroy adversary space capability. Battle management and command and control provide the ability to integrate space control with other joint force activities in the prosecution of warfighting. Obviously, multiple electronic, directed-energy, or kinetic system applications and concepts can be envisioned for effectively engaging in space control. Platform basing can be an issue when such concepts are being considered, and from this perspective the potential advantages offered by sea-based platforms appear reasonably significant.

Security classification restrictions prohibit meaningful discussions in this report of all space control efforts—specific plans, shortfalls, and technology gaps are largely classified. Suffice it to say that it would be in the best interests of the United States to pursue broad "disrupt, degrade, deny, or destroy" countermeasure capabilities to apply against any space-based capabilities that contribute to the threat posed by potential adversaries. Although the Under Secretary of the Air Force, as the DOD Executive Agent for Space, has the lead for space control efforts, it appears that other Services could support these activities within their own areas of competence. Thus, it would appear appropriate for the Navy to pursue potential sea-based space control concepts in close coordination with the Air Force.

Recommendation 4.22. The Department of the Navy should explore potential sea-based space control concepts in coordination with the activities of the DOD Executive Agent for Space.

5

Fulfilling Naval Forces Space Needs: A Vision

The Department of the Navy must fully support and exploit the ongoing transformation of the Department of Defense (DOD) and intelligence community. This transformation will maximize the integration and use of modern communications, sensing, and knowledge technology. The resultant changes will be revolutionary in enabling U.S. forces to fight more effectively, more efficiently, and with ever more exacting precision. The major investments (measured in tens of billions of dollars) to be made over the next 10 years by the DOD and the intelligence community in fielding the next generation of communications, surveillance, intelligence, and knowledge systems will (if properly executed and exploited by the Services) have a profound effect on U.S. forces. Within this transformation, space will play an increasingly important role. The previous chapters illustrate that the Navy's current capstone concept Sea Power 21 will directly depend on space support—thus, it is critical for the Department of the Navy to treat space support as a priority. The Committee on the Navy's Needs in Space for Providing Future Capabilities believes that if the Department of the Navy implements the recommendations made in this study, the Navy and Marine Corps will be better able to transform successfully into an even more effective and relevant force within the DOD.

The committee believes that the Department of the Navy, the Chief of Naval Operations, and the Commandant of the Marine Corps need to create a vision for Navy space systems and the role of space in future naval systems. The committee proposes a vision in which the Navy and Marine Corps will be positioned to specify, develop, and utilize space systems when these systems provide essential, cost-effective capabilities that achieve the objectives of Sea Power 21.

To accomplish the objectives of this vision, the Department of the Navy will need to continue to use a variety of methods to assure that operational space-based capabilities are available to and used by naval forces. The Navy will be able to assure that space capabilities are provided to the warfighter through a combination of approaches. As detailed throughout this report, these approaches include the following:

- Increased participation by the Department of the Navy in National Security Space (NSS) organizations;
- Continued support by Navy leadership of the acquisition and operation of Navy-specific space systems, such as the Mobile User Objective System (MUOS) and Geodetic Satellite (Geosat) Follow-on;
- Continued Navy utilization of NSS systems, assured through the Navy's resource contributions to NSS programs and the NSS planning and budgeting process;
- Continued execution and support of Navy and Marine Corps efforts through the Technical Exploitation of National Capabilities Program (TENCAP) to augment existing space-based capabilities;
- Increased participation in the planning, development, and acquisition of space systems developed by other agencies, including those of the National Aeronautics and Space Administration, the National Oceanic and Atmospheric Administration, and the Defense Advanced Research Projects Agency;
- Improved naval space planning, programming, and budgeting activities through a renewed focus on systems analysis, modeling, and simulation to support effective development of space requirements as described in more detail in Chapter 3 of this report;
- Increased and redirected investments in Navy space science and technology (S&T) and experimentation in order to better address prioritized Sea Power 21 capability gaps, consistent with leveraging the S&T and experimentation investments of NSS organizations and other agencies; and
- Establishment of a partnership role for the Navy in all NSS organizations, following the National Reconnaissance Office (NRO) model for Navy participation, and thus the achievement of expanded roles for the naval space cadre and expanded interaction in NSS activities.

To successfully execute these elements of the committee's vision for meeting the Navy forces' needs in space, the Navy should accelerate the development of and expand the naval space cadre to assure that the Navy has sufficient qualified personnel in these areas:

- *NSS organizations*—to fulfill Department of the Navy requirements (detailed in DOD Directive 5101.2[1]) to participate in the planning, programming, and acquisition activities of the DOD Executive Agent for Space;

---

[1]Paul Wolfowitz, Deputy Secretary of Defense. 2003. "DOD Executive Agent for Space," DOD Directive 5101.2, Department of Defense, Washington, D.C., June 3. The complete text of this directive is presented in Appendix B of this report.

• *Fleet Forces Command, Marine Corps Headquarters, and the Office of the Chief of Naval Operations (in particular, the Office of the Deputy Chief of Naval Operations for Warfare Requirements and Programs)*—to develop space-related strategy and operations and associated naval requirements and concepts for space systems, space doctrine, education and training requirements, and standards for space research, development, testing, evaluation, and acquisition;

• *The Office of the Chief of Naval Operations (in particular, the Office of the Deputy Chief of Naval Operations for Resources, Requirements, and Assessments)*—to develop Navy-specific space strategies, plans, and programming information for coordination with the DOD Executive Agent for Space, integration into the NSS programs, and to support DOD-wide planning, programming, and budgeting;

• *The operational forces (Navy and Marine Corps)*—to provide educated recommendations from operators on naval, DOD, and NSS program investments and priorities to support naval warfighters; and

• *The Space and Naval Warfare Systems Command, the Office of Naval Research, and Defense Advanced Research Projects Agency*—to lead naval-specific research, development, and acquisition programs to fill naval requirements.

Niels Bohr once said, "Prediction is extremely difficult, especially about the future." Given the difficulties of prediction, the committee does offer a vision for the future of naval forces engaged in the development and exploitation of space-based capabilities. This vision is illustrated by the following "snapshots," highlighting the kinds of possibilities that could lie ahead.

Snapshot 1. A joint experiment program including the Naval Center for Space Technology (NCST), the Air Force Space Battlelab, and the Space Test Program validates the fusion of data from a new NCST-designed and -built space-based hyperspectral sensor with airborne active light detection and ranging (LIDAR) sensor data to demonstrate unparalleled characterization of the littorals.

Snapshot 2. A team of technologists from the Navy Warfare Development Command (NWDC)—deployed with the fleet as part of an annual program to gain a better understanding of current Sea Power 21 capability gaps—identifies an opportunity to provide wideband communications to small ships by adapting an Army-developed Aerostat to solve the antenna-placement and field-of-view challenges facing small ships.

Snapshot 3. The new, airborne moving target indication (AMTI) spacecraft development program under the NSS is headed by a Navy captain and member of the Navy space cadre. The systems engineering team is led by an Air Force lieutenant colonel who received his Ph.D. in space network engineering from the Naval Postgraduate School and is assisted by a civilian aerospace engineer detailed from the Naval Center for Space Technology.

Snapshot 4. Major John Smith, USMC, commanding officer of Company D, 4th Marine Expeditionary Battalion and graduate of the Naval Postgraduate School's space operations curriculum, develops the operational needs statement for a hypersonic suborbital vehicle to get a quick-response antiterrorism team from the Marine Corps Base at Quantico, Virginia, to a classified, Third World location to recover a nuclear warhead from a terrorist training base.

Snapshot 5. The new Chief of Naval Research is a member of the Navy space cadre and has had a tour as a program director at DARPA.

Snapshot 6. Rear Admiral J.P. Jones, USN, Director of Requirements at the Air Force Space Command, is selected for his third star and assignment as commander of the U.S. Seventh Fleet.

The Navy is at an important juncture, with the DOD and the intelligence community transforming themselves to leverage and fully exploit the technical revolutions in communications, sensors, and knowledge systems. Space systems are at the forefront of this transformation, enabling unprecedented improvements in connectivity, efficiency, and precision. With the Department of the Navy's adoption of the recommendations articulated in this report, its acceptance of the challenges presented, and its provision of personal leadership, the Naval Services will be able to achieve a vision of their future role in space.

Appendixes

A

Department of the Navy History in Space

EARLY DEVELOPMENTS

As a result of World War II, the United States, its allies, and its adversaries realized a number of profound technological capabilities (nuclear weapons, radar, electronic navigation, weapon guidance, long-range rockets (V-2), proximity fuzes, and so on) that would affect warfare forever. Yet by modern standards, the Department of Defense (DOD) in general and the Navy in particular had many deficiencies, including the following:

- Long-haul wireless communications were limited to the high-frequency (HF) band and were often not available as a result of little-understood changes in the environment;
- Navigation was inaccurate and uncertain—even when the Navy's Long Range Navigation (LORAN) system was available, inaccuracies were generally in the range of 1 to 2 miles;
- Environmental knowledge (regarding winds, wave height, cloud cover, storms, temperature, and sea conditions) was limited to the local area of an observer, and forecasting capabilities were limited or nonexistent;
- Except for HF transmissions, the ability to track and identify beyond-line-of-sight (BLOS) targets or transmitters did not exist;
- Weapon delivery accuracy was appallingly poor, being limited by the lack of precise knowledge of the geolocations of both the weapon release platform and the target;
- Target surveillance and identification were limited to the questionable capabilities of reconnaissance aircraft whose survival over enemy terrain was tenuous;

- Surface-to-surface rockets had a maximum range of about 200 miles and an apogee of about 60 miles, used single-stage nongimbaled engines, delivered a unitary payload, and were highly unreliable in their performance; and
- The ability to identify and locate the site of a clandestine detonation of a nuclear weapon was rudimentary.

Few knowledgeable military officers of the post-World War II era would have disagreed with the foregoing list of shortfalls in military capabilities. While the senior leadership of the Navy had no game plan for overcoming these deficiencies, there was a faith, best expressed in Vannevar Bush's book *Endless Horizons*,[1] that broad investment in basic research would generate discoveries that would lead to the development of new technologies to ameliorate this list of deficiencies.

The Navy, more than any other Service, embarked on a systematic program of investment in and support of basic research in its own in-house laboratories (primarily the Naval Research Laboratory (NRL)), in universities, in university-associated contract research centers (such as the Applied Physics Laboratory (APL) at Johns Hopkins University, the Applied Research Laboratory at Pennsylvania State University, and the Scripps Institution of Oceanography), and in selected industrial research centers.

Modern space-based capabilities have largely served to resolve the list of post-World War II deficiencies listed above. Almost all of the modern space-based capabilities available to the DOD and the Navy are traceable to early investments in basic research made by the Navy and other DOD Services in the decades after the war.

Many vectors drove NRL's interests in the use of space platforms. The Navy came out of World War II with communications systems that were dependent on the vagaries of HF propagation. As early as 1927, NRL developed an intense interest in attaining an understanding of the factors that described the behavior of the ionosphere. By 1937, NRL had established a small group to develop sounding rockets that could carry research instruments above the atmosphere. Much of that work was quiescent, but not forgotten, during the war years.

At the end of World War II, the U.S. Army captured a factory near Niedersachswerfen in Germany containing enough parts of the Vergeltungswaffe (the vengeance weapon, or V-2) to allow the reconstruction of about 100 V-2 rockets. These were shipped to the Army's White Sands Missile Range (WSMR) in New Mexico where the Army had the mission of developing long-range tactical missiles. If the Army was going to fly these missiles in space, it needed to obtain a better understanding of the environment above the atmosphere. By early 1946, NRL had established a group with responsibility to investigate the physical phenomena in, and the properties of, the upper atmosphere with a view to supplying

[1]Vannevar Bush. 1975. *Endless Horizons*, Public Affairs Press, Washington, D.C.

knowledge to influence the course of future military operations. As NRL was the most competent group in the country in the area of above-atmosphere instrumentation, the Army group at WSMR turned to NRL for support, and a synergistic union between the two resulted.

Although the V-2 was an unreliable rocket, it could lift scientific payloads of 500 kg approximately 150 km above Earth. NRL teams led by Herbert Friedman and Richard Tousey were able to undertake landmark research that established the effect of solar x-rays on the ionosphere and established the nature of the solar ultraviolet spectra.

Through the late 1940s and early years of the 1950s, the NRL-WSMR team continued its work with ever more impressive results. However, it was clear that the supply of V-2 rockets was dwindling. Beginning in the late 1940s, personnel in NRL's Rocket Sonde Research Branch began the design of a new rocket designed to support NRL's upper-atmosphere research program. NRL's first rocket was called Viking. Unlike the V-2, which was not steerable, Viking had a gimbaled engine, allowing it to be steered. After the normal early developmental problems, Viking became routinely available for research purposes.

Although the Viking could deliver a scientific payload to greater altitudes than the V-2 could, its real advantage was its steerable engine. The jump from a single-stage steerable rocket to a multistage steerable rocket was, conceptually at least, a straightforward engineering challenge. Once a multistage-steerable-rocket capability was available, the placement of a satellite in orbit would be possible. By 1954, as a result of developments at NRL and at the Army's Redstone Arsenal in Alabama, the conceptual pathway to placing research packages into satellite orbits was clear.

In the summer of 1954, members of the International Scientific Radio Union recommended that as part of the activities scheduled for the International Geophysical Year (IGY)—1957/1958—artificial satellites be launched for use as research platforms. On July 29, 1955, the White House announced that, in support of the IGY, the United States intended to launch "small earth circling satellites."[2] At the time of this announcement, a high-level committee known as the Steward Committee was convened by the Assistant Secretary of Defense. The Steward Committee was charged with deciding who in fact would be assigned responsibility for the first satellite launch.

The Air Force might have been a contender for this assignment, but its major personnel talent was committed to the development of the Atlas rocket, which was slated to become the launch vehicle for an intercontinental ballistic missile and, by congressional guidance, had to take precedence over support of a scien-

---

[2]James C. Hagerty, Press Secretary to the President. 1955. "Presidential Press Briefing," The White House, Washington, D.C., July 29. Available online at <http://www.eisenhower.archives.gov/dl/IGY/StatementbyHagertyJuly291955.pdf> or at <http://www.hq.nasa.gov/office/pao/History/sputnik/17.html>. Accessed May 4, 2004.

tific rocket project. That left two available rocket design groups: one was at NRL and the other, led by Wernher von Braun, was at the Army's Redstone Arsenal.

The Army proposed the Orbiter, some of whose components already had been under development with funds provided by the Office of Naval Research (ONR). Plans for the Orbiter called for a multistage rocket that could put a 5 lb satellite into orbit. The NRL proposal was for a three-stage rocket using a Viking-based design for the first two stages and a newly designed upper stage that could deliver a 40 lb scientific payload into orbit. The NRL proposal also included a virtually ready-to-go radar tracking system, called Minitrack, that already had been developed to provide tracking for the Viking project.

After considerable internal debate, the Steward Committee selected the NRL approach, and on September 9, 1955, NRL won stewardship of the satellite program, which became known as Project Vanguard. NRL's basic tasking was to build a satellite launch vehicle,[3] place one satellite in orbit, verify its orbital path, and accomplish one scientific objective, all before the end of the IGY.

After suffering numerous highly visible failures (in part in an effort to catch up with the Soviet launch of Sputnik), the NRL team on March 17, 1958, delivered a 3½ lb satellite into an orbital trajectory. Although Project Vanguard suffered every public relations calamity conceivable, it represented an astounding achievement. In just 2½ years, NRL took an all-paper design to a successfully launched satellite. The legacy of the Project Vanguard rocket design would be traceable through several National Aeronautics and Space Administration (NASA) and Air Force vehicles, including the Delta, on which Vanguard is the second stage.

The activities of this period eventually brought the United States into the era of manned spaceflight. The Minitrack system would become the basis of all DOD satellite tracking, including the advanced and comprehensive Naval Space Surveillance System (NAVSPASUR). (See Box A.1 for additional detail on NRL's development of satellite tracking systems.) The existence of NAVSPASUR also proved crucial to the development and monitoring of the Global Positioning System (GPS) constellation. In addition to these accomplishments, the Vanguard I satellite provided data that enabled improved calculations of Earth's shape and of the periodic variations of the density of the upper atmosphere.

Several weeks after the Vanguard launch, President Eisenhower set in motion actions that would lead to the formation of NASA. Personnel from NRL, the Army Ballistic Missile Agency, the Air Force Cambridge Research Laboratory, and the Jet Propulsion Laboratory were transferred into the nascent NASA organization. A draft of NRL personnel with experience in rocket design constituted the largest portion of NASA's initial technical staff.

---

[3]Rocket development was performed in close partnership with the Martin Company of Baltimore, a predecessor of Lockheed Martin Corporation.

BOX A.1
Early Development of Satellite Tracking

Surprising the nation in the late 1950s was the Soviet Union's launch, on October 4, 1957, of the Sputnik satellite. The U.S. Navy, through the almost immediate response of the Naval Research Laboratory (NRL), started receiving data and tracking Sputnik by its third orbit, fewer than 5 hours after its launch, using the NRL's radio array at Hybla Valley, Virginia. NRL soon brought its Vanguard Minitrack satellite tracking system online and used it to provide additional higher-precision tracking data for Sputnik.[1] Sputnik orbital measurements were also collected by the Applied Physics Laboratory of Johns Hopkins University and led to the use of measurements of the Doppler shift of Sputnik's radio signal, enabling a greatly improved tracking calculation. The use of Doppler-shift monitoring and tracking later led to the concept of the Transit series of navigation satellites.[2]

---

[1]Louis Gebhard. 1979. *Evolution of Naval Radio-Electronics and Contributions of the Naval Research Laboratory*, Report 8300, Naval Research Laboratory, Washington, D.C.

[2]William K. Klingaman. 1993. *APL—Fifty Years of Service to the Nation*, Johns Hopkins University/Applied Physics Laboratory, Laurel, Md., pp. 119-122.

Although the formation of NASA left NRL without in-house rocket design capabilities, NRL still had a highly competent satellite design group and a strong research staff with impressive qualifications for undertaking exo-atmospheric research programs. By 1962, NRL demonstrated the launch of multiple satellites with a single rocket. After this seminal event, the launches of classified satellites often were piggybacked on launches of NRL research satellites. One particularly successful program developed electronic-intelligence-gathering satellites (called GRAB, for Galactic Radiation and Background), as described in Box A.2. These

BOX A.2
Navy Electronic Intelligence Satellite Development

The nation's first successful electronic intelligence (ELINT) satellite was proposed by the Naval Research Laboratory (NRL) in the spring of 1958. Named GRAB, for Galactic Radiation and Background, its first launch was approved by President Eisenhower in May 1960, four days after a Central Intelligence Agency U-2 aircraft was lost on a reconnaissance mission over Soviet territory—thus initiating an urgent need to develop continual unmanned surveillance from space. Taking advantage of NRL's multiple-launch capability, the GRAB satellite launch also carried the Navy's third Transit satellite and the Navy's SOLRAD satellite—developed to measure solar radiation.

launches were so successful that during the 1960s and 1970s classified satellite programs and associated staff grew to the point of dominating NRL's space activities.

By 1960, Sputnik, Vanguard, and several other satellites had been placed in orbit. Although much of the necessary technology had not yet matured sufficiently to allow the implementation of space systems, a broad general understanding of their future role had begun to evolve in the DOD, in the intelligence community, in the nation's science community, and in many civil agencies within the U.S. government. This time period witnessed the development of many new space capabilities, including the following:

- The ability, demonstrated by NRL, to communicate with in-flight sounding rockets established that communications satellites would be feasible.
- Navy satellite communications firsts were achieved, including the first shore-to-shore (1960),[4] ship-to-shore (1963),[5] and ship-to-aircraft (1963)[6] satellite-based communications links.
- Imagery retrieved from NRL cameras launched on pre-1960 sounding rockets indicated the potential for satellite imagery.
- Early NRL satellite launches demonstrated the feasibility of launching multiple independent packages from a single launch—indicating the potential for developing intercontinental ballistic missiles with multiply targeted independent reentry vehicles.
- Although the Vanguard satellite was ridiculed for its small size and multiple launch failures before finally being placed in orbit, it provided a vast amount of data on the actual shape of Earth. These data were in turn used to achieve a significant improvement in the delivery accuracy of intercontinental ballistic missiles.
- Early NRL efforts on the readout of telemetry and sensor data indicated that if space-based sensors were specifically configured to intercept signals from hostile radars or communications transmitters, tremendous increases in intelligence-gathering capabilities and emitter-location capabilities could be achieved. Thus, the limitations of surface- and aircraft-based sensors could be eliminated.
- The satellite tracking networks developed in support of Project Vanguard and the development under ONR sponsorship of reliable, stable, and highly accurate atomic clocks and hydrogen masers implied that if such clocks could be

[4]Louis Gebhard. 1979. *Evolution of Naval Radio-Electronics and Contributions of the Naval Research Laboratory*, Report 8300, Naval Research Laboratory, Washington, D.C.

[5]CDR D.J. Woodward, USN, and CDR J.F. Debold, USN. 1965. "Kingsport Experience with the Syncoms," *Astronautics and Aeronautics*, January, pp. 30-37.

[6]RADM B.F. Roeder, USN. 1964. "Naval Communications of the Future," *Signal*, May, pp. 62-64.

space-qualified, ultraprecise space-based navigation systems (such as GPS) might ultimately be deployed.

- The performance of Earth-orientated sensors in early NRL and APL tests, and particularly by the Television and Infrared Observation Satellite (TIROS) funded by the Advanced Research Projects Agency (ARPA), indicated that weather and sea-surface conditions could be observed worldwide with considerable accuracy and precision and that space-sensor-derived data could provide significantly improved weather and ocean forecasting capabilities. These activities were first consolidated into NASA and eventually became one of the core operations of the National Oceanic and Atmospheric Administration (NOAA).
- Sensors used in early NRL sounding rocket experiments to detect x-rays and solar radiation were also used to validate that space-based sensors could provide an improved capability to globally detect nuclear explosions. The Defense Advanced Research Projects Agency (DARPA) and NRL later collaborated on the early deployment of the Vela Hotel satellites,[7] whose measurements enabled the U.S. government to determine that there was a high probability that clandestine nuclear tests in space could be detected and thus that the United States could verify compliance with a nuclear test ban treaty with the Soviet Union.

During the decades after 1960, there was an explosive growth in the overall national funding for space-related systems. As computer capabilities, sensor performance, clock accuracies, optical system resolution, communications capabilities, and the thrust of launch vehicles grew, so did the performance of systems that were launched. From the standpoint of senior naval officials, the performance of such systems was impressive. Yet, faced with the problems of conducting the Vietnam War and competing with the Soviet Union's development of an open-ocean surface and submarine fleet, the performance of space systems in the 1960s through the 1980s was not great enough to command a significant priority in the Navy's budgetary allocations.

INTELLIGENCE, SURVEILLANCE, AND RECONNAISSANCE

When it was available, imagery derived from space-based sensors was quickly found to be of tactical use to the Services. Unfortunately, when imagery was not available for technical or priority reasons or both, it often took weeks to months before Navy or other Service requests for imagery were satisfied. Satellite imagery was the domain of the intelligence community, whose highest priorities in those decades were driven by considerations of nuclear exchanges between

[7]The Vela Hotel series was developed to detect nuclear explosions in space and has become one of America's most successful military space programs.

the United States and the Soviet Union. In that period the support of military operations definitely had low priority in the eyes of those who controlled the tasking of imaging satellites. As a consequence, Services including the Navy had little desire to provide budgetary (or even personnel) support for such activities.

A similar problem existed with respect to satellites that were designed either to intercept hostile communications or to geolocate radars associated with hostile surface-to-air missiles (broadly classified as electronic intelligence (ELINT) systems). Although the computers of that era could not perform near-real-time analyses of the data, the real delay was inherent in the operation of the era's ELINT satellites. At the time, all satellite-derived intelligence data were forwarded to the National Security Agency (NSA) where they were processed and evaluated by NSA's analytical staff. Data that could be associated with nuclear warfare had first priority; data related to the support of military operations were a distant second priority.

Until the early 1980s, the latency of space-derived ELINT data could be days to weeks. As a consequence, the Navy's concepts of operations did not have a strong dependence on the availability of such data. Although both NRL and certain naval commands were heavily involved in space-based programs related to the gathering of ELINT from space-based sensors, the Navy as a whole was not sufficiently impressed with system output to be willing to provide major personnel or budgetary support for these programs. Nonetheless, NRL continued to develop space-based intelligence systems relying on funding supplied from outside the Navy (see Box A.3).

The Navy's limited involvement in such programs had a negative feedback effect. Owing in part to the long lead times for developing new satellite systems (often 5 to 15 years), as new generations of equipment were designed and procured the Navy's small amount of up-front monetary and personnel support led to a lack of early priority for Navy and naval requirements. In addition, the less responsive the acquisition community became to Navy requirements, the less interest the Navy showed in the development and use of future systems.

ENVIRONMENTAL SENSING AND GEODESY

The first movie of a hurricane taken from overhead was recorded in 1954 by NRL.[8] Since that time, the Navy has developed a significant number of scientific and environmental sensors and satellites, including the following: SOLRAD, the Solar Radiation satellite; Geosat, the Navy's Geodetic Satellite; and, most recently, Coriolis—launched through the Space Test Program (STP), this satellite carries the NRL-built WindSat wind-speed and wind-direction measurement sen-

[8] Louis Gebhard. 1979. *Evolution of Naval Radio-Electronics and Contributions of the Naval Research Laboratory*, Report 8300, Naval Research Laboratory, Washington, D.C., p. 397.

BOX A.3
Intelligence Satellite Development by the Naval Research Laboratory

In a sense, the Naval Research Laboratory (NRL) became a victim of its own success. Beginning in 1960, the nation's investment in space-based systems expanded tremendously. The expansion was far beyond anything that an organization even as large as NRL could cope with. Although NRL developed and launched some 80 satellites in the four decades between 1960 and 2000, the number was only 1 or 2 percent of the total national effort. NRL's expertise in sensor technology and scientific research caused the expertise of its staff to be sought out and funded by a long list of non-Navy organizations (the Air Force, Army, Missile Defense Agency, National Aeronautics and Space Administration, National Oceanic and Atmospheric Administration, National Security Agency, National Reconnaissance Office (NRO), National Science Foundation, predecessors of the National Geospatial-Intelligence Agency, and others). The support that NRL received from all of these non-Navy organizations was sufficient to allow it to build up a large and very competent staff.

Through the early 1980s, NRL's largest effort was in support of the development of classified prototype surveillance satellites. These developments were indeed successful, to the degree that the sponsoring agency (NRO) decided to go into serial production of such satellites. Since NRL was a research laboratory and not a manufacturing facility, the production of the next generation of satellites was transferred to an industrial organization. This transfer left many members of the NRL staff without sponsor support and necessitated a rather traumatic drawdown in the number of NRL personnel available to manage the development, acquisition, and launch of full satellite systems. The difficulty was that the Navy (through the Office of Naval Research) only provided funds to cover NRL's basic research activities. In the past three decades, no Navy funds have been provided to NRL to develop and launch new satellite systems. As a consequence, NRL's ability to develop and deploy satellite systems that offer the Navy new warfighting capabilities has diminished by a significant amount.

sor.[9] The operation of this latter sensor system is coordinated with NASA, NOAA, and the Air Force. SOLRAD was used to provide measurements of the Sun to help with predicting radio transmission performance and to help understand other scientific issues. The concepts underlying Geosat and its forerunners began after the Navy's Moon Bounce program showed that accurate ranging data (in this case the Earth-Moon distance) could be measured. This led NRL scientists to determine that similar radar, looking at the ocean surface from orbit, could accu-

[9]The WindSat payload is also a risk-reduction demonstration for NOAA's National Polar-orbiting Operational Environmental Satellite System.

rately determine the ocean surface height.[10] Satellites supporting these altimetry measurements included Seasat, which carried a radar altimeter built by APL, and the current Geosat system, also built by APL. These altimetry satellites provide insights into ocean surface topography, currents, and eddies that are important for submarine-launched ballistic missiles and other submarine operations, landings, and Special Forces operations.

The Defense Meteorological Satellite Program (DMSP) carries NRL's Special Sensing Microwave/Imager (SSM/I) sensors and the follow-on SSM/IS, which measures sea ice, precipitation, atmospheric winds, and surface winds. The current generation of surface-monitoring sensors, the WindSat sensor built by NRL, adds to the prior measurements by providing data on wind speed and direction. These wind data have proven important to many naval operations and are now included in plans for NOAA's next generation of polar-orbiting environmental satellites.

Early geodetic work (modeling of Earth's surface height and gravitational field) was done by both the Army Map Service and the U.S. Naval Observatory using precision measurements of the motion of Earth's natural satellite, the Moon. This approach, while improving geodesy significantly over earlier terrestrial techniques, still left surface-height uncertainties on the order of kilometers. With the advent of satellites orbiting close to Earth (hundreds of miles in altitude), more and higher-precision geodesy data were collected. The APL Transit satellite, the first satellite designed to provide accurate positioning data, enabled significant increases in geodetic accuracy, eventually reducing the uncertainties to the 10 m range.[11] Later, launch of the GPS satellites, with their resident high-precision clocks and advanced ground-monitoring systems, brought geodetic measurement accuracies on the order of millimeters, enabling small ground movements such as continental drift and small movements along faults to be measured. These data, in combination with laser altimeter data provided by the Geosat satellites, enable accurate calculation of Earth's surface profile as well as its gravitational field profile. The gravimetric data in particular have been useful to the Navy in support of the submarine-launched ballistic missile system.

From the 1960s through the 1980s, the Navy expressed intense interest in the acquisition of satellites to provide environmental data. In this area, NOAA, NASA, and the Air Force had requirements congruent with those of the Navy, and to the degree that contemporary technology permitted, many but not all Navy needs were satisfied by the satellites acquired and operated by NOAA. To meet the Services' special environmental sensing needs, particularly to determine

---

[10]B.S. Yapee, A. Shapiro, D.L. Hammond, B.D. Au, and E.A. Uliana. 1971. "Nanosecond Radar Observation of Ocean Surface from a Stable Platform," *IEEE Transactions on Geoscience Electronics*, GE-9(3), pp. 170-174.

[11]Robert J. Danchik. 1998. "An Overview of Transit Development," *Johns Hopkins APL Technical Digest*, January-March, Vol. 19, No. 1, p. 24.

cloud-free areas for optical imaging satellites, the DMSP was established under the direction of the Air Force. Thus, there was no strong need for additional Navy funding in the area of space systems that supported the needs of the weather forecasting community.

The "O" in NOAA stands for Oceanic, but it is the "A" standing for Atmospheric that has historically dominated the organization's activities. Early environmental satellites, fielded to provide worldwide synoptic information on sea-surface temperatures and radar images of the sea surface, were developed and supported by NASA. Although the data from NASA oceanographic satellites were designed for the national scientific community, the Navy was a massive consumer and user of the derived information. Further advances in the performance of oceanographic satellite sensors will require space-based active sources (radar and lasers) to measure ocean wave heights and to measure near-surface winds and humidity. Neither NASA nor NOAA has shown recent budgetary enthusiasm for investments in active space-based environmental sensors. Since 1994 NOAA has been the designated National Executive Agent for environmental satellites;[12] thus, the Navy can only submit requests to the interagency steering group of the National Polar-orbiting Operational Environmental Satellite System (NPOESS) satellites to bring about what the Naval Oceanographic Office believes to be necessary and desirable improvements in the performance of active space-based environmental sensors.

SATELLITE COMMUNICATIONS

Beginning in the 1960s, the Navy, along with the rest of the DOD and the commercial communications industry, was intensely interested in the development of communications satellites whose potential had been demonstrated by early NRL rocket studies. In the following decades, the DOD's investment in satellite communications systems was large, but it was dwarfed by worldwide commercial investments and developments in this area. Although the Navy's investments in this same area were robust, they tended to be specific to Navy needs for systems with antennas suitable for shipboard use. These activities included support (and even acquisition) of several satellite communications systems, such as the Fleet Satellite (FLTSAT) communications system, UHF (ultra-high frequency) Follow-on (UFO), and the extremely high frequency (EHF) packages on UFO.

The Navy recognized fairly early on that satellite communications to submarines would require the use of very small antennas. This need, as well as others, led to the Navy's support for the development of EHF-band communications satellites. The Navy worked with the Massachusetts Institute of Technology Lin-

[12]National Science and Technology Council. 1994. "Convergence of the U.S.-Polar-Orbiting Operation Environmental Satellite Systems," Presidential Decision Directive, May.

coln Laboratory in the 1960s to explore the feasibility of providing EHF satellite communications; this work led to the development and demonstration of the Lincoln Experimental Satellite series.

In 1974 the DOD allowed the Air Force to fund the entire development of a radiation-hardened DOD EHF communications satellite constellation (later named MILSTAR, for Military Strategic, Tactical, and Relay Satellite) in exchange for the Air Force's being designated the lead Service for EHF communications. Even though the Air Force was developing MILSTAR, the Navy, through the Lincoln Laboratory, developed and deployed Fleet EHF Packages (FEPs) as test packages on the final four FLTSAT satellites as well as on the initial four UFO satellites.[13] This action assured the Navy of timely access to EHF communications without reliance on the developing MILSTAR program.

Just after the 1970 removal of DOD restrictions against Navy development of operational satellite systems, the Navy obtained DOD authorization to develop FLTSAT to provide the fleet with global support for tactical communications. FLTSAT satellites were initially procured in 1972 through the Air Force, with the NRL developing and providing the Fleet Broadcast Processor (FBP)—a system that provided a large margin of jam resistance and one that continues to be used on the UFO and other UHF satellites.[14]

As a result of the uncertainties and delays in the acquisition of the FLTSAT satellites, the Navy was permitted, beginning in 1976, to lease UHF satellite communications from commercial sources. The first FLTSAT satellite was finally launched in 1978. As other Services and agencies began using FLTSAT and its reliability was proven, the Navy sought and received permission to purchase additional satellites on a "turnkey" basis.[15] This program, called Leasat, entered operation in 1984. Based on the successes and demonstrated economy of these UHF programs, the Navy was assigned the responsibility for acquiring UFO as well as for developing the next-generation UHF system, the Mobile User Objective System (MUOS).

By the 1990s the DOD recognized that Service-unique satellite communications systems were undesirable. At a minimum, the DOD decided that all space communications systems needed to be joint and to satisfy the communications needs of all Services. Although the Navy has continued to invest resources in the development of MUOS and other satellite communications systems, these efforts are no longer dominated by Navy needs and requirements.

---

[13]Gary Federici, Robert Hess, and Kent Pelot. 1997. *From the Sea to the Stars: A History of U.S. Navy Space and Space-Related Activities*, Working Paper, The Center for Naval Analyses, Alexandria, Va., pp. 101-102.

[14]Gary Federici, Robert Hess, and Kent Pelot. 1997. *From the Sea to the Stars: A History of U.S. Navy Space and Space-Related Activities*, Working Paper, The Center for Naval Analyses, Alexandria, Va., p. 69.

[15]Whereby the Navy accepts and pays for a satellite after the contractor has the satellite on-orbit and can show that the satellite is functioning within design specifications.

POSITION, NAVIGATION, AND TIMING

In the area of space-based navigational systems, the development of the necessary technology took place before senior Navy leadership could appreciate its significance. Leading these navigation efforts, ARPA provided funding for the first Navy satellite navigation system, Transit (further described in Box A.4). The

BOX A.4
The First U.S. Satellite Navigation System: Transit

The Navy's ballistic missile submarines had a critical need for precise location updates in order to meet their operational goals. The Navy, through the Applied Physics Laboratory (APL) at Johns Hopkins University, invented and developed Transit—the world's first satellite navigation system—to meet that need.

There had been no navigation program at APL before the initiation of Transit. The Transit concept was developed after APL tracked the Soviet Sputnik "surprise" of October 4, 1957. The tracking of the Sputnik satellite by William Guier, a researcher at APL, led him to recognize that the measured Doppler shift of a satellite's radio signal could be used to accurately track the satellite. Guier and George Weiffenbach, also at APL, tracked Sputnik by this method and improved the orbit predictions over the next 6 months. This capability led Frank McClure, then also at APL, to recognize that "you could use the Doppler effect to compute your location on the ground."[1]

On March 18, 1958, McClure's concept was described in detail in a memorandum.[2] The Transit concept was briefed to a special subcommittee of the presidential science advisory committee headed by Herbert York, who had visited APL earlier. York later became head of the Advanced Research Projects Agency (ARPA) and approved ARPA's providing initial funding for Transit. Later funding was provided by the Navy: the Transit program was then funded and managed through the Navy's Strategic Programs Office, allowing it to proceed with a minimum of red tape. "There were no constraints put upon us," observed Weiffenbach. "The only question I ever heard from them was, 'do you need more money?'"

On September 17, 1959, the first Transit satellite was launched. Unfortunately it did not achieve orbit, owing to a booster malfunction. The second launch, on April 13, 1960, was successful, and it demonstrated the key features needed for an operational system. After 10 launches of progressively refined prototype satellites, the Transit satellite system became operational in 1964 and remained so until 1996. This system provided accurate navigational information throughout the Cold War years to the most covert component of the ballistic missile triad, submarines. Transit went on to be used on a large number of naval and commercial ships as well as extensive numbers of land-based systems.[3]

---

[1]William K. Klingaman. 1993. *APL—Fifty Years of Service to the Nation*, Johns Hopkins University/Applied Physics Laboratory, Laurel, Md., p. 122.

[2]F.T. McClure to R.E. Gibson. 1958. APL Memorandum dated March 18, 1958.

[3]See Wiliam K. Klingaman, 1993, *APL—Fifty Years of Service to the Nation*, Johns Hopkins University/Applied Physics Laboratory, Laurel, Md., pp 105-129, for additional background on Transit.

capabilities of the present-day GPS system are an outgrowth of NRL work in the 1960s and early 1970s on Transit and Timation (TIMe/navigATION), the first satellite to utilize on-orbit atomic clocks, and NRL's continuing support to NAVSPASUR; Aerospace Corporation's work on the downlink signal structure; ONR's continued support of the development of highly stable atomic clocks that were made space-qualified by NRL; and the microelectronics revolution that allowed the design of compact GPS receivers. These and other Navy efforts in support of GPS are listed in Box A.5.

In the 1970s the Navy had neither GPS guided weapons nor plans to acquire such capabilities. In effect, the Navy withdrew its support of further development

BOX A.5
Contributions by the Naval Research Laboratory to the Global Positioning System

The Naval Research Laboratory (NRL), mainly through the efforts of a talented radio engineer, Roger Easton, developed and demonstrated the core concepts of the current Global Positioning System (GPS). Based on NRL's efforts, a low-cost, low-risk path to the operational capability for GPS was provided. The key features of NRL's efforts were as follows:

- The concept of GPS;[1]
- The technical results of four generations of navigation satellites launched and operated from 1967 through the launch of the second Navigation Technology Satellite (NTS-2) in 1977, the first GPS satellite, and 7 follow-on years of operations support;[2]
- The geodetic Earth model optimized for the GPS orbit and geodetic-quality satellite tracking systems;[3]
- Operational precise orbit predictions;[4]
- Ground positioning accuracies meeting the joint requirements, proven using the NRL-developed Timation system;

---

[1]Naval Research Laboratory (NRL). 1964. "A Satellite Navigation System," NRL Space Surveillance Branch, Washington, D.C. Technical Memorandum #112, June 9.

[2]A very fundamental part of this activity was verifying Einstein's relativistic clock shift. The offset was refined by the NTS-2 satellite to within 1.5 percent of Einstein's theory and is still used today in the operational GPS satellites. This effect was one of the key reasons for NRL's adopting circular orbits for Timation, so as to enable the satellites to stay in an approximately constant gravitational field.

[3]NRL developed and the Naval Surface Warfare Center (NSWC) deployed a tracking network of five ground stations to gather precise GPS tracking data (needed to refine the GPS orbital geodetic information) as well as to monitor each GPS clock.

of space-based navigational capabilities and the development of the GPS system was assigned by the DOD to the Air Force and its lead laboratory at Aerospace Corporation. NRL and the Navy retained some responsibility for the continued development of ever more precise space-qualified clocks that will be inserted into future generations of GPS satellites.

SPACE SURVEILLANCE

As described above, NRL early on proved the effectiveness of its Minitrack satellite tracking system. For many years Minitrack was also used by NASA as its

- High-precision satellite clocks, for which NRL continues to provide the necessary on-orbit precision clock monitoring;[5]
- The Atlas F as a low-cost launch capability;
- The Vandenberg Air Force Base Satellite Launch Complex Three West, used for NRL's NTS-2 and Rockwell's National Development Satellite (NDS) launches;
- Extensive supporting technologies tested in the midaltitude space environment, including nickel-hydrogen batteries and new classes of solar cells; and
- Orbit injection and operations for Timation and NTS continuing through 1984, provided at the Navy's Blossom Point Tracking Facility near La Plata, Maryland.

In addition, NRL engineers provided the Air Force and its contractors, Aerospace Corporation and Rockwell International, a detailed NTS-2 operations plan as a means of ensuring that those involved in the developing Air Force GPS program understood the designs, expected performance, and operation of NTS-2.[6] NRL then engaged frequently with the Rockwell and Air Force engineers during GPS's early operational phases. The launch in 1978 of the first NDS satellite, by Rockwell, demonstrated that a successful transition of skills and information had taken place.

---

[4]The NRL and NSWC ground stations reported orbital and clock data, as did the Air Force tracking sites when they became operational, to NSWC for processing with the geodetic and other data described elsewhere. The NSWC provided highly precise ephemeris sets and trajectories for many years. These were used by NRL in all of its precision navigation work, and later these data were utilized at the GPS Master Control Station at Vandenberg Air Force Base.

[5]The measured clock data are analyzed at NRL and posted as part of the Frequency Standards Working Group. The data are currently on a protected NRL Web site, with data available on every operational navigation satellite dating back to the launch of first Timation satellite in 1967.

[6]Rockwell, Air Force, and Naval Research Laboratory. 1976. NTS-2 Operations Plan, January 27.

primary satellite tracking system. The Transit Network (TRAnet) and Operational Network (OPnet) were developed to support the tracking and monitoring needs of Transit and other satellites and remained in use into the 1990s.

The NAVSPASUR system, developed by NRL, has been a key to the nation's ability to track almost all objects in Earth orbit that are larger then about 10 cm in diameter. Being an active radar system (one currently based on radio-frequency transmit and receive arrays positioned in a line across the entire southern United States), it detects all orbital objects, not only active (transmitting) satellites. NAVSPASUR was initiated in 1958 as Minitrack and was commissioned as an operational naval command in February 1961. Data processing is done, as it has been since the early 1960s, at the Space Surveillance Processing Center in Dahlgren, Virginia.[16] This system has been very productive in keeping the United States aware of almost all large objects orbiting Earth, producing an average of 160,000 observations per day.[17] NAVSPASUR is still highly effective, and while responsibility for the system was transferred to the Air Force in 2003, the Air Force has retained the Navy's Dahlgren facility to continue its support of NAVSPASUR operation.

One additional ramification of NAVSPASUR was that, with the addition in 1960 of the second radio array site in Texas, the NAVSPASUR system was found to be in need of a means to coordinate the high-precision clocks on which the facilities rely. This timing coordination need then led to the concept of a space-based common time and became one of the drivers for the Navy's Timation program.[18]

---

[16]Gary R. Wagner. 2004. "Navy Passes Down the Space Watch," *NNSOC Domain Magazine*, Winter, pp. 4-5.

[17]Gary R. Wagner. 2004. "The Building of a National Treasure," *NNSOC Domain Magazine*, Winter, pp. 6-7.

[18]Ronald L. Beard, James A. Buisson, and Roger L. Easton, Naval Research Laboratory. To be published. *From Vanguard to GPS: The Role of the Naval Research Laboratory in the Development of the Global Positioning System, 1955-2000*, GPS History Version 8.0 draft report, Naval Research Laboratory, Washington, D.C., pp. 7-8.

B

Department of Defense Directive 5101.2

Department of Defense

DIRECTIVE

NUMBER 5101.2
June 3, 2003
Certified Current as of November 21, 2003

Incorporating Change 1, July 22, 2003
DA&M

SUBJECT: DOD Executive Agent for Space

References: (a) Section 113 of title 10, United States Code
(b) DOD Directive 3100.10, "Space Policy," July 9, 1999
(c) DOD Directive 5101.1, "DOD Executive Agent," September 3, 2002
(d) Secretary of Defense Memorandum, "National Security Space Management and Organization," October 18, 2001
(e) through (m), see enclosure 1

1. PURPOSE

Pursuant to the authority vested in the Secretary of Defense under reference (a), and consistent with the policies in reference (b), this Directive:

1.1. Establishes policy and assigns responsibilities and authorities for the planning, programming, and acquisition of space systems within the Department of Defense (DOD).

1.2. Designates the Secretary of the Air Force as the DOD Executive Agent for Space in accordance with reference (c).

1.3. Implements and supersedes paragraph 3.2. of reference (d), consistent with reference (e), and supersedes reference (f).

2. APPLICABILITY

This Directive applies to the Office of the Secretary of Defense; the Military Departments; the Chairman of the Joint Chiefs of Staff; the Combatant Commands; the Office of the Inspector General, Department of Defense; the Defense Agencies; the DOD Field Activities; and all other organizational entities in the Department of Defense (hereafter collectively referred to as "the DOD Components").

3. MISSION

The DOD Executive Agent for Space, exercising the responsibilities and authorities herein, shall develop, coordinate, and integrate plans and programs for space systems and the acquisition of DOD space Major Defense Acquisition Programs to provide operational space force capabilities to ensure the United States has the space power to achieve its national security objectives.

4. DEFINITIONS

As used in this Directive, the following terms have the meaning set forth below:

4.1. Space Forces. The space and terrestrial systems, equipment, facilities, organizations, and personnel necessary to access, use, and, if directed, control space for national security.

4.2. Space Power. The total strength of a nation's capabilities to conduct and influence activities to, in, through, and from the space medium to achieve its objectives.

4.3. Space Systems. All of the devices and organizations forming the space network. These consist of: spacecraft; mission package(s); ground stations; data links among spacecraft, ground stations, mission or user terminals, which may include initial reception, processing, and exploitation; launch systems; and directly related supporting infrastructure, including space surveillance and battle management/command, control, communications, and computers.

4.4. Ballistic Missile Defense System. For the purposes of this Directive, the Ballistic Missile Defense System is not considered a space Major Defense Acquisition Program in accordance with the Secretary of Defense memorandum, dated January 2, 2002 and the Under Secretary of Defense for Acquisition, Technology, and Logistics (USD(AT&L)) memorandum, dated February 13, 2002 (references (g) and (h)).

5. POLICY

It is DOD policy that:

5.1. Planning and programming of space systems and acquisition of DOD space Major Defense Acquisition Programs shall be performed on a DOD-wide basis by a DOD Executive Agent for Space, in accordance with applicable law and pertinent Federal and DOD policies and regulations.

5.2. The product of DOD space planning shall be documented in the National Security Space Plan (NSSP) based on annually updated space plans and architectures of the DOD Components developed in coordination with the Deputy Director of Central Intelligence for Community Management (DDCI/CM) and other Federal officials, as appropriate.

5.3. The Program Objective Memoranda of the DOD Components shall be reviewed annually by the Under Secretary of the Air Force-Director of the National Reconnaissance Office (USecAF-DNRO), who shall prepare the National Security Space Program Assessment (NSSP A), which reports on the consistency of the implementation of defense and intelligence space programs with policy, strategy, planning, and programming guidance, and architectural decisions, based on the Future Years Defense Program.

5.4. Interservice support agreements, memoranda of understanding, and other written arrangements, shall be used to the maximum extent practicable, in accordance with reference (e), to establish the support arrangements necessary for the DOD Executive Agent for Space to accomplish the mission and fulfill the responsibilities assigned by this Directive.

5.5. The Heads of DOD Components shall ensure the full cooperation of personnel under their jurisdiction with the DOD Executive Agent for Space to enable the Department to eliminate duplication of effort and ensure that DOD-wide planning, programming, space research, development and acquisition resources are used effectively.

6. RESPONSIBILITIES

6.1. The Secretary of the Air Force is hereby designated as the DOD Executive Agent for Space and in that role shall:

6.1.1. Exercise DOD-wide responsibilities for planning and programming of space systems and acquisition of DOD space Major Defense Acquisition Programs.

6.1.2. Redelegate the DOD Executive Agent for Space responsibilities only to the Under Secretary of the Air Force who also serves as the DNRO. No further delegation is authorized.

6.2. The DOD Executive Agent for Space shall:

6.2.1. Ensure all DOD Executive Agent for Space responsibilities and functions are assigned and executed, in accordance with reference (c) and this Directive.

6.2.2. Recommend to the Heads of the DOD Components DOD-wide processes for the development, coordination, integration, review, and implementation of space system plans, strategy, and acquisition programs.

6.2.3. Represent and advocate DOD-wide space interests in the planning and programming processes and defense acquisition process. Provide information to OSD Principal Staff Assistants (PSAs) as requested to support program analysis and evaluation of space policy, plans, and acquisition programs. Provide information copies of such submissions to the USD(AT&L), the Under Secretary of Defense for Policy (USD(P)), the Under Secretary of Defense for Intelligence (USD(I)), the Assistant Secretary of Defense for Networks and Information Integration (ASD(NII)), the Director, Program Analysis and Evaluation (P A&E), and the Chairman of the Joint Chiefs of Staff.

6.2.4. Integrate the needs and requirements of the DOD Components into space plans and major space program requirements documents. Resolve issues with the DOD Components and then submit architectures and requirements to the Joint Requirements Oversight Council (JROC) for validation. Adjudicate unresolved requirements and interoperability issues through the JROC. Provide space plans to the JROC for information.

6.2.5. Prepare the annual NSSP in consultation with the Heads of DOD Components and the DDCI/CM. Provide the NSSP to the USD(AT&L), the USD(P), the Under Secretary of Defense (Comptroller) (USD(C)), the USD(I), the ASD(NII), the Director, PA&E, the Chairman of the Joint Chiefs of Staff, and the DDCI/CM.

6.2.6. Recommend DOD-wide space and space-related planning and programming guidance to the USD(P) and the Director, Program Analysis and Evaluation (P A&E) for consideration in their formulation of planning and programming guidance documents. Provide information copies of such submissions to the USD(AT&L), the ASD(NII), and the Heads of DOD Components.

6.2.7. Consult with the USD(AT&L), the USD(I), the ASD(NII), and the Vice Chairman of the Joint Chiefs of Staff on significant changes to any space program subject to Major Defense Acquisition Program review before such changes are included in a Program Objective Memorandum submission of a DOD Component.

6.2.8. Develop and recommend to the USD(AT&L), in coordination with the Heads of the DOD Components, policies and programs that improve, streamline, and strengthen DOD Component space and space-related technology access and development programs; encourage space-related open market competition and technology-driven prototype efforts that offer increased military capabilities at lower total ownership costs and faster fielding times; and exploit the cost-reduction potential of accessing innovative or commercially developed technologies.

6.2.9. When serving as the Milestone Decision Authority for DOD space Major Defense Acquisition Programs in accordance with the USD(AT&L) memoranda, dated February 14, 2002 and October 2, 2002 (references (i) and (j)), or such subsequent delegation, supervise the execution of DOD space Major Defense Acquisition Programs.

6.2.9.1. Inform the USD(AT&L) of all waivers and exemptions to DOD policy granted for DOD space programs, and of all new or revised acquisition processes established for DOD space Major Defense Acquisition Programs.

6.2.9.2. Provide recommendations to the USD(AT&L) for all certifications, reports, and waivers for DOD space Major Defense Acquisition Programs required by Chapter 144 of title 10, United States Code.

6.2.10. Develop assessments and, where appropriate, recommend policies and strategies to the USD(AT&L) to maintain the capability of the U.S. space industry to meet DOD needs.

6.2.11. Develop space systems acquisition plans, strategies, guidance, and assessments to ensure that acquisition milestone review and the Planning, Programming, and Budgeting System processes are timely and are implemented effectively.

6.2.12. Ensure that the DOD Component space Major Defense Acquisition Programs are carried out through joint or integrated program offices to the maximum extent practicable. These program offices shall report to the USecAF through their respective Component Acquisition Executive.

6.2.13. Develop and implement a process to align Air Force and NRO programs. Encourage each organization to use the "best practices" found throughout the national security space community for space research, development, acquisition, and operations. Such actions shall be coordinated with the USD(AT&L), the USD(P), the IUSD(C), the USD(I), the ASD(NII), the Director, PA&E, the Chairman of the Joint Chiefs of Staff, the Director for Operational Test and Evaluation (DOT&E), and the DDCI/CM.

6.2.14. Subject to the authority of the DOT&E, oversee development, test, and evaluation planning and operational test and evaluation of DOD space Major Defense Acquisition Programs and other space programs on the OSD Test and Evaluation Master Oversight List. For any such program being acquired, the acquiring Department shall be responsible for the test and evaluation and the provision of pertinent documentation in a timely manner to the DOD Executive Agent for Space.

6.2.15. Review annually, in coordination with the other DOD Components and the Intelligence Community, the space program, budget, and accounting mechanism (referred to as a "virtual" Major Force Program for Space) established by the Director, PA&E and the DDCI/CM, and recommend to the Director, PA&E changes to the content of the "virtual" Major Force Program for Space.

6.2.16. Serve as the Air Force Acquisition Executive for Space.

6.2.17. Submit Selected Acquisition Reports to the USD(AT&L).

6.2.18. Establish, maintain, and preserve records that document the transaction of business and mission of the DOD Executive Agent for Space to provide evidence of the organization, functions, policies, procedures, decisions, and activities in accordance with DOD Directive 5015.2 (reference (k)).

6.3. The Heads of the DOD Components shall:

6.3.1. Participate in the planning, programming, and acquisition activities of the DOD Executive Agent for Space. In support of these activities, provide to the DOD Executive Agent for Space information regarding:

6.3.1.1. Space science and technology priorities, programs, and funding.

6.3.1.2. Operational requirements for space and space-related systems. The Missile Defense Agency shall provide a Description of Operational Capability in lieu of operational requirements in accordance with references (f) and (g).

6.3.1.3. Strategies, plans for space systems, force structure, capabilities, measures of performance, and schedules.

6.3.1.4. Approved programmatic and budget data for space programs.

6.3.1.5. Space acquisition program data.

6.3.1.6. Key indicators reflecting the status of, or changes to, their cadre of space professionals.

6.3.1.7. Readiness of space forces when there are implications that may assist the DOD Executive Agent for Space in addressing a space system deficiency.

6.3.1.8. Recommendations on priorities for space support to the DOD Components.

6.3.1.9. Operations Concepts for space and space-related systems.

6.3.2. Provide space strategies, plans, and program information to the DOD Executive Agent for Space for review, coordination, and integration into the NSSP and to support DOD-wide planning, programming, and acquisition. Space program information normally shall be provided throughout the year and will include a Program Objective Memoranda submission or such other mutually agreeable mechanisms as established by the DOD Executive Agent for Space and the Heads of the DOD Components.

6.3.3. Submit space needs and requirements to the DOD Executive Agent for Space for integration into space plans and major space program requirements documents as well as associated acquisition programs prior to submitting requirements to the JROC. Where possible, resolve issues with the DOD Executive Agent for Space. Unresolved requirements and interoperability issues shall be adjudicated by the JROC.

6.3.4. Develop DOD Component requirements and concepts for: space systems; space doctrine, education, and training requirements and standards; space research, development, testing, evaluation, and acquisition; related military construction; and space-related strategy and operations. In coordination with the appropriate DOD Components, provide such information to the DOD Executive Agent for Space. Where appropriate, use established DOD processes for the development of joint doctrine, training, and strategies.

6.3.5. Develop and maintain a sufficient cadre of space-qualified personnel to support their Component in space planning, programming, acquisition, and operations. Support the DOD Executive Agent for Space with space cadre personnel to represent their Component in DOD-wide planning, programming, and acquisition activities.

6.3.6. Recommend space and space-related planning and programming guidance for DOD Component programs to the USD(P) and the Director, PA&E for consideration in their formulation of planning and programming guidance documents. Inform the USecAF, the USD(AT&L), the USD(I), and the ASD(NII) of such submissions.

6.3.7. Continue to develop, acquire, and fund space research, development, and acquisition programs that meet DOD Component requirements and submit such program information to the DOD Executive Agent for Space in accordance with this Directive.

6.3.8. Advise the DOD Executive Agent for Space on Program Objective Memoranda that significantly change any program subject to review during the program assessment for space, before submission to the OSD.

6.4. The OSD Principal Staff Assistants shall, within their functional areas, exercise their designated authorities and responsibilities as established by law, Executive Order, or DOD guidance to facilitate the mission of the DOD Executive Agent for Space and to implement their respective actions as specified herein and required by reference (d).

7. RELATIONSHIPS

7.1. The DOD Components shall implement their space responsibilities, functions, and authorities in accordance with this Directive. Nothing herein shall be interpreted to subsume or replace the functions, responsibilities, or authorities of the OSD PSAs or the Heads of DOD Components, prescribed by law, Executive Order, or DOD guidance.

7.2. In performing assigned responsibilities, the DOD Executive Agent for Space shall:

7.2.1. Report to the Secretary and Deputy Secretary of Defense and be subject to the authority, direction, and control of the Secretary of Defense in accordance with reference (a).

7.2.2. Maintain close communications with the Heads of DOD Components and OSD PSAs responsible for space and space-related matters. Matters pertaining to space plans, programming, and acquisition programs shall be appropriately coordinated with these officials to ensure their requirements and equities for space systems and capabilities are met.

7.2.3. Coordinate with the Heads of DOD Components on their space program science and technology, research, development, and production programs to eliminate duplication of effort and ensure that resources are used to achieve maximum effect. Such actions shall be coordinated with the USD(AT&L) and the Director, Defense Research and Engineering.

7.2.4. Support the USD(P) and the Director, PA&E, as requested, with program analysis and evaluation of space policy, plans, and programs. Provide information on such support to the USD(AT&L), the USD(I), and the ASD(NII).

7.2.5. Support the USD(C) with the development, integration, implementation, and maintenance of space program financial strategic plans as well as the reengineering of associated business practices.

7.2.6. Support the USD(C), as requested, in the preparation and validation of economic analyses in support of space program financial systems.

7.2.7. Support the National Security Space Architect (NSSA) in the development, coordination, and integration of space architectures to achieve efficiencies in acquisition and future operations.

7.2.8. Assist in the development of the annual NSSPA conducted by the NSSA in consultation with the Heads of the DOD Components, the DDCI/CM, and other Federal officials. The USecAF-DNRO shall submit the NSSPA through the Secretary of the Air Force to the Senior Executive Council established by DOD Directive 5105.66 (reference (1)) and shall provide the NSSPA to the

USD(AT&L), the USD(P), the USD(C), the USD(I), the Director, PA&E, and the ASD(NII). In coordination with the DDCI/CM, the USecAF-DNRO shall provide the NSSPA for review jointly to the Secretary of Defense and the Director of Central Intelligence.

8. AUTHORITY

8.1. Obtain reports and information consistent with DOD Directive 8910.1 (reference (m)), as necessary, to carry out assigned functions.

8.2. Communicate directly with the Heads of the DOD Components, as necessary to carry out assigned functions, to include the transmission of requests for advice and assistance. Communications to the Military Departments shall be transmitted through the Secretaries of the Military Departments, their designees, or as otherwise provided in law or directed by the Secretary of Defense in other DOD issuances. Communications to the Commanders of the Combatant Commands shall be transmitted through the Chairman of the Joint Chiefs of Staff.

9. EFFECTIVE DATE

This Directive is effective immediately.

Paul Wolfowitz
Deputy Secretary of Defense

Enclosures—1
E 1. References, continued

El. ENCLOSURE 1

REFERENCES, continued

(e) DOD Instruction 4000.19, "Interservice and Intragovernmental Support," August 9, 1995
(f) DOD Directive 5160.32, "Development of Space Systems," September 8, 1970 (hereby canceled)

(g) Secretary of Defense Memorandum, "Missile Defense Program Direction," January 2, 2002
(h) Under Secretary of Defense for Acquisition, Technology, and Logistics Memorandum, "Ballistic Missile Defense Program Implementation Guidance," February 13, 2002
(i) Under Secretary of Defense for Acquisition, Technology, and Logistics Memorandum, "Delegation of Milestone Decision Authority for DOD Space Systems," February 14, 2002
(j) Under Secretary of Defense for Acquisition, Technology, and Logistics Memorandum, "Independent Cost Estimates for Space Programs," October 2, 2002
(k) DOD Directive 5015.2, "DOD Records Management Program," March 6, 2000
(1) DOD Directive 5105.66, "Senior Executive Council," July 10, 2001
(m) DOD Directive 8910.1, "Management and Control of Information Requirements," June 11, 1993

C

Sea Power 21 Capability Areas: Assessment of Dependence on Space Mission Areas

TABLE C.1 Assessment of Current Sea Strike Capability Dependency on the Intelligence, Surveillance, and Reconnaissance (ISR), Meteorology and Oceanography (METOC), and Position, Navigation, and Timing (PNT) Space Mission Areas

Capability	Projected ISR Dependency	Current Space Support	Lead Agency	Comments/ Limitations	Projected METOC Dependency	Current Space Support
Strike	Critical	Useful		Target identification, location, and tracking; decoy discrimination		
Conduct strike operations	Critical	Useful			Critical	Effective
—Engage fixed land targets	Critical	Effective	NRO, USAF	Persistency, area coverage		
—Engage moving targets	Critical	Limited	USAF	GMTI required; area coverage, ocean coverage, low-latency communications		
Conduct special operations	Critical	Useful				
—Provide precision targeting	Critical	Useful	NRO, USAF	Persistency, FIA improvements late, SBR/GMTI		
—Conduct direct action	Critical	Useful	NRO	Persistency	Critical	Useful

Lead Agency	Comments/ Limitations	Space PNT Dependency	Current Space Support	Lead Agency	Comments/ Limitations
	Carrier operations, weather prediction, mission planning	Critical	Effective	USAF, NAVAIR, USNO	Weapons system integration, limited resistance to interference, time synchronization
NOAA	Inconsistent reliability, availability and latency of data	Critical	Effective		
		Critical	Effective		
		Critical	Limited		
		Critical	Effective		
		Critical	Effective		
USN, NOAA, Commercial, NTM	Tactical decision aids; littoral characterization. Latency, tasking, spectral resolution, spatial resolution, only passive systems via NOAA executive agency, no transition path from advanced technology development to operations.	Critical	Effective		

continues

TABLE C.1 Continued

Capability	Projected ISR Dependency	Current Space Support	Lead Agency	Comments/ Limitations	Projected METOC Dependency	Current Space Support
Conduct offensive information operations	Limited	Limited				
—Jam potential threats	Limited	Limited		Better done from air or ground		
—Conduct network attack	Limited	Limited		Better done from air or ground		
Provide aircraft survivability	Critical	Useful	NRO	Tactical ELINT, and operational ELINT improvements		
Naval Fire Support	Limited	Limited				
Provide preparation fires	Limited	Limited			Contributory	Useful
Provide close supporting fires	Limited	Limited				
Provide precision supporting fires	Limited	Limited				
Provide volume fires	Limited	Limited				
Ship to Objective Maneuver	Critical	Useful				

Lead Agency	Comments/ Limitations	Space PNT Dependency	Current Space Support	Lead Agency	Comments/ Limitations
		Critical	Effective		
		Limited	Effective		
		Critical	Effective		
		Critical	Effective		
		Critical	Effective	USAF, SPAWAR, USNO	Integrated GPS, time synchronization
NOAA, USAF	International ownership, reliability, availability, and latency	Limited	Effective		
		Limited	Effective		
		Critical	Effective		
		Limited	Effective		
		Critical	Effective	USAF, SPAWAR, USNO	Limited resistance to interference

continues

TABLE C.1 Continued

Capability	Projected ISR Dependency	Current Space Support	Lead Agency	Comments/ Limitations	Projected METOC Dependency	Current Space Support
Project forces, reposition forces	Critical	Useful	USAF, NRO	Persistence, area coverage, ocean coverage	Critical	Useful
Assault centers of gravity and critical vulnerabilities	Critical	Useful	USAF, NRO	Persistence, area coverage, ocean coverage	Limited	
Conduct concurrent follow-on missions	Critical	Useful	USAF, NRO	Persistence, area coverage, ocean coverage, battle damage assessment	Critical	Useful
Strategic Deterrence	Critical	Limited				
Conduct nuclear strike	Critical	Useful	USAF, NRO	Targeting, reconnaissance, area coverage		

Lead Agency	Comments/ Limitations	Space PNT Dependency	Current Space Support	Lead Agency	Comments/ Limitations
USN, NOAA, commercial, NTM	Trafficability, littoral needs. Latency, tasking, spectral resolution, spatial resolution, only passive systems via NOAA executive agency, no transition path from advanced technology development to operations.	Critical	Effective		
		Critical	Effective		
Commercial, NTM	Battle damage assessment, trafficability, nowcasting. Latency, tasking, spectral resolution, spatial resolution, no transition path from advanced technology development to operations. No hyperspectral or SBR support.	Critical	Effective		
		Critical	Effective	USAF, USN SSPO, USNO	Antijam communications, timing synchronization
		Critical	Effective		

continues

TABLE C.1 Continued

Capability	Projected ISR Dependency	Current Space Support	Lead Agency	Comments/ Limitations	Projected METOC Dependency	Current Space Support
Provide assured survivability	Critical	Limited	USAF, NRO	Theater AMTI, BMD, terrorist		

NOTES: Acronyms are listed in Appendix G. "Critical" indicates a Sea Power 21 capability strongly dependent on the indicated space mission area. "Contributory" indicates a Sea Power 21 capability whose performance can be enhanced by access to the indicated space mission area. "Effective" indicates that Navy support to link the space and Sea Power 21 capabilities is at an appropriate level to meet Navy needs. "Useful" indicates that Navy support to link the space and Sea Power 21 capabilities is helping support Navy needs, but could see improvement. "Limited" indicates little or no connection between capability areas or little Navy interaction to link the capability areas.

Lead Agency	Comments/ Limitations	Space PNT Dependency	Current Space Support	Lead Agency	Comments/ Limitations
		Critical	Limited		Limited HEMP protection and resistance to interference

TABLE C.2 Assessment of Current Sea Shield Capability Dependency on the Intelligence, Surveillance, and Reconnaissance (ISR), Meteorology and Oceanography (METOC), and Position, Navigation, and Timing (PNT) Space Mission Areas

Capability	Potential ISR Dependency	Current Space Support	Executive Agency	Comments/ Limitations	Potential METOC Dependency	Current Space Support
Force Protection	Contrib-utory	Limited				
Protect against SOF and terrorist threats	Critical	Useful	NRO, USAF	Persistence, GMTI, AMTI, low cross-section targets, area coverage	Contributory	Limited
Mitigate effects of CBRNE	Limited	Limited	NRO	Hyperspectral		
Surface Warfare	Critical	Useful				
Neutralize near-horizon surface threats	Contrib-utory	Useful	NRO, USAF	ELINT, IMINT: FIA, SBR/GMTI, no AMTI	Limited	Limited
Neutralize OTH surface threats	Critical	Useful	NRO, USAF	ELINT, IMINT: FIA, SBR/GMTI, no AMTI	Limited	Limited
Undersea Warfare	Contrib-utory	Limited				
Provide self-defense against subsurface threats	Contrib-utory	Limited	NRO, USAF	Persistence, ocean coverage, area coverage		

Executive Agency	Comments/ Limitations	Space PNT Dependency	Current Space Support	Executive Agency	Comments/ Limitations
		Limited	Effective	USAF, USNO, NAVSEA	Limited resistance to interference
USN, USAF, NTM	Wave height, currents, surface temperature; ELINT, spatial resolution	Limited	Effective		
		Limited	Limited		
		Critical	Effective	USAF, USNO, NAVAIR	
NOAA	Atmospheric ducting, mission can be accomplished using onboard sensors	Limited	Effective		
NOAA	Atmospheric ducting	Critical	Effective		
		Critical	Effective	USAF, USNO, SYSCOMs	Persistence
		Limited	Limited		

continues

TABLE C.2 Continued

Capability	Potential ISR Dependency	Current Space Support	Executive Agency	Comments/ Limitations	Potential METOC Dependency	Current Space Support
Neutralize submarine threats in the littorals	Contrib-utory	Limited	NRO, USAF	Persistence, ocean coverage	Critical	Limited
Neutralize mines in shallow to deep water	Contrib-utory	Limited	NRO, USAF	Persistence, ocean coverage, area coverage	Critical	Useful
Breach surf zone, minefields, obstacles, and barriers	Contrib-utory	Limited	NRO, USAF	Persistence, ocean coverage	Critical	Limited
Theater Air and Missile Defense	Critical	Limited				
Provide self-defense against air and missile threats	Critical	Limited	USAF	SBIRS-H, no planned AMTI		
Provide maritime air and missile defense	Critical	Limited	USAF	SBIRS-H, no planned AMTI		

Executive Agency	Comments/ Limitations	Space PNT Dependency	Current Space Support	Executive Agency	Comments/ Limitations
NTM, commercial	Currents, surf, sediment, turbidity, bioluminescence, bathymetry; spatial and temporal resolution, spectral range/resolution	Critical	Limited		
USN, USAF, NOAA	Wave height, currents, surface temperature; spatial resolution	Limited	Limited		
NTM, commercial, NTM	Wave height, currents, surface temperature; water-penetrating spectral sensors, active microwave; no passive or active hyperspectral, no SBR, poor littoral temporal and spatial resolution	Critical	Limited		
		Critical	Limited	USAF, MDA, USNO	Limited resistance to interference, LPI and LPD communications, time synchronization
		Critical	Limited		
		Critical	Limited		

continues

TABLE C.2 Continued

Capability	Potential ISR Dependency	Current Space Support	Executive Agency	Comments/ Limitations	Potential METOC Dependency	Current Space Support
Provide overland air and missile defense	Critical	Limited	USAF	SBIRS-H, no planned AMTI		
Maintain sea-based NRO missile defense	Critical	Limited	USAF	SBIRS-H, no planned AMTI		

NOTES: Acronyms are listed in Appendix G. "Critical" indicates a Sea Power 21 capability strongly dependent on the indicated space mission area. "Contributory" indicates a Sea Power 21 capability whose performance can be enhanced by access to the indicated space mission area. "Effective" indicates that Navy support to link the space and Sea Power 21 capabilities is at an appropriate level to meet Navy needs. "Useful" indicates that Navy support to link the space and Sea Power 21 capabilities is helping support Navy needs, but could see improvement. "Limited" indicates little or no connection between capability areas or little Navy interaction to link the capability areas.

Executive Agency	Comments/ Limitations	Space PNT Dependency	Current Space Support	Executive Agency	Comments/ Limitations
		Critical	Limited		
		Critical	Limited		

TABLE C.3 Assessment of Current Sea Basing Capability Dependency on the Intelligence, Surveillance, and Reconnaissance (ISR), Meteorology and Oceanography (METOC), and Position, Navigation, and Timing (PNT) Space Mission Areas

Capability	Potential ISR Dependency	Current Space Support	Lead Agency	Comments/ Limitations	Potential METOC Dependency	Current Space Support
Deploy and Employ	Contrib-utory	Useful				
Close the force and maintain mobility	Contrib-utory	Useful	NRO, USAF	Persistence, area coverage, ocean coverage		
Provide at-sea arrival and assembly	Contrib-utory	Useful	NRO, USAF	Persistence, area coverage, ocean coverage	Critical	Effective
Allow selective offload	Limited	Limited				
Reconstitute and regenerate at sea	Limited	Limited				
Provide Integrated Joint Logistics	Limited	Limited				
Provide sustainment for operations at sea	Limited	Limited				
Provide sustainment for operations ashore	Limited	Limited				
Provide focused logistics	Limited	Limited				

Lead Agency	Comments/ Limitations	Space PNT Dependency	Current Space Support	Lead Agency	Comments/ Limitations
		Limited	Effective	USAF	Global coverage, persistence
		Limited	Effective		
NOAA, USAF	Ocean routing; existing infrastructure is working; space and time resolution gradually improving.	Limited	Effective		
		Critical	Effective		
		Limited	Effective	USAF, DLA	Global coverage, persistence
		Limited	Effective		
		Limited	Effective		
		Limited	Effective		

continues

TABLE C.3 Continued

Capability	Potential ISR Dependency	Current Space Support	Lead Agency	Comments/ Limitations	Potential METOC Dependency	Current Space Support
Provide shipboard and mobile maintenance	Limited	Limited				
Provide force medical services	Limited	Limited				
Provide advanced base support	Limited	Limited				
Pre-position Joint Assets Afloat	Limited	Limited			Critical	Effective
Integrate and support joint personnel and equipment	Limited	Limited				
Provide afloat command and control physical infrastructure	Contrib- utory	Useful	NRO, USN	Space ISR processing, fusion, exploitation		
Provide AFSB capability for joint operations	Limited	Limited				

NOTES: Acronyms are listed in Appendix G. "Critical" indicates a Sea Power 21 capability strongly dependent on the indicated space mission area. "Contributory" indicates a Sea Power 21 capability whose performance can be enhanced by access to the indicated space mission area. "Effective" indicates that Navy support to link the space and Sea Power 21 capabilities is at an appropriate level to meet Navy needs. "Useful" indicates that Navy support to link the space and Sea Power 21 capabilities is helping support Navy needs, but could see improvement. "Limited" indicates little or no connection between capability areas or little Navy interaction to link the capability areas.

Lead Agency	Comments/ Limitations	Space PNT Dependency	Current Space Support	Lead Agency	Comments/ Limitations
		Limited	Effective		
		Limited	Effective		
		Limited	Effective		
NOAA, USAF	Ocean routing; existing infrastructure is working; space and time resolution gradually improving.	Critical	Effective	USAF	Global GPS coverage needed, limited signal resistance to interference
		Limited	Effective		
		Critical	Effective		
		Critical	Effective		

TABLE C.4 Assessment of Current FORCEnet Capability Dependency on the Intelligence, Surveillance, and Reconnaissance (ISR) and Position, Navigation, and Timing (PNT) Space Mission Areas

Capability	Potential ISR Dependency	Current Space Support	Lead Agency
Communications and Data Networks	Limited	Limited	
Provide communications infrastructure	Limited	Limited	
Provide network protection	Contributory	Limited	
Provide network synchronization	Limited	Limited	
Provide information transfer	Limited	Limited	
Intelligence, Surveillance, and Reconnaissance	Critical	Useful	
Conduct sensor management and information processing	Critical	Useful	NRO, USAF
Detect and identify targets	Critical	Useful	NRO, USAF
—Fixed land targets	Critical	Useful	NRO, USAF
—Moving land targets	Critical	Limited	USAF
—Air and missile targets	Critical	Limited	USAF
—Surface targets	Critical	Limited	NRO, USAF
—Submarine targets	Contributory	Limited	USAF, NRO

Comments/Limitations	Space PNT Dependency	Current Space Support	Lead Agency	Comments/Limitations
	Critical	Limited	USAF, DISA, SPAWAR	Global coverage, persistence, limited resistance to interference
	Critical	Limited		
Monitor/warn of threats to nodes and communications	Critical	Limited		
	Critical	Effective		
	Critical	Limited		
	Critical	Effective	USAF, NGA	Global coverage, persistence, time synchronization
Space sensor ISR tasking, processing, fusion, exploitation; SBR, FIA, SBIRS	Critical	Effective		
Space sensor ISR tasking, processing, fusion, exploitation; SBR, FIA, SBIRS	Critical	Limited		
Persistence, FIA, SBR/SAR	Critical	Effective		
Persistence, SBR/GMTI	Critical	Limited		
AMTI, SBIRS	Critical	Effective		
ELINT, FIA, SBR/SAR	Critical	Effective		
Persistence, area coverage, ocean coverage	Critical	Limited		

continues

TABLE C.4 Continued

Capability	Potential ISR Dependency	Current Space Support	Lead Agency
—Mines	Contributory	Limited	USAF, NRO
Provide cueing and targeting information	Critical	Useful	NRO, USAF
Assess engagement results	Critical	Useful	NRO, USAF
Common Operational and Tactical Picture	Critical	Useful	
Provide mission planning	Critical	Useful	NRO
Provide battle management synchronization	Limited	Limited	
Provide common PNT and environmental information	Limited	Limited	
Integrate and distribute sensor information	Contributory	Useful	NRO, USAF
Track and facilitate engagement of time-sensitive targets	Critical	Limited	NRO, USAF
Track and facilitate engagement of non-time-sensitive targets	Critical	Useful	NRO, USAF

NOTES: Acronyms are listed in Appendix G. "Critical" indicates a Sea Power 21 capability strongly dependent on the indicated space mission area. "Contributory" indicates a Sea Power 21 capability whose performance can be enhanced by access to the indicated space mission area. "Effective" indicates that Navy support to link the space and Sea Power 21 capabilities is at an appropriate level to meet Navy needs. "Useful" indicates that Navy support to link the space and Sea Power 21 capabilities is helping support Navy needs, but could see improvement. "Limited" indicates little or no connection between capability areas or little Navy interaction to link the capability areas.

Comments/Limitations	Space PNT Dependency	Current Space Support	Lead Agency	Comments/Limitations
Persistence, area coverage, ocean coverage	Critical	Limited		
ELINT, FIA, SBR/GMTI/SAR; persistence, area coverage, ocean coverage	Critical	Effective		
BDA, IMINT, ELINT; SBR, MTI, SAR	Critical	Effective		
	Critical	Effective	USAF, USN	Time synchronization, limited resistance to interference
BDA, IMINT, ELINT; SBR, MTI, SAR	Critical	Effective		
	Critical	Effective		
	Critical	Effective		
Space ISR fusion with theater and force assets	Critical	Limited		
DSP only today: future SBIRS, SBR/GMTI, FIA; no planned ATMI; persistence, area and ocean coverage	Critical	Limited		
DSP, ELINT, IMINT today: future SBIRS, SBR/GMTI, FIA; no planned ATMI	Critical	Effective		

D

Space Communications Systems and Capabilities

The Navy and Marine Corps currently employ a wide range of space communications services and will employ new Department of Defense (DOD) satellite communications systems as they come online. Some of these systems are geared primarily for voice (walkie-talkie) service, others for Internet Protocol (IP) datagrams; some are protected and robust against jamming, others are very fragile; and some are designed for handheld devices, others for high-speed data to large-deck ships. As the wide range of these systems can be very confusing, the following discussion summarizes the basic structure of DOD space communications strategy, in the context of space communications programs. Military space communications can be categorized in four areas on the basis of communications systems requirements: wideband services, narrowband services, assured secure communications services, and commercial communications services.

Figure D.1 provides a notional mapping onto user classes of the newer DOD satellite systems, circa 1999, when the new satellite systems were being defined. Reading down the "Navy Ships" column, one observes that the Advanced Extremely High Frequency (AEHF) expanded data rate (XDR), Wideband Gapfiller System (WGS) Ka- and X-band, and UHF (ultrahigh frequency) Follow-on (UFO) services were considered most relevant for the surface Navy, with the Global Broadcast Service (GBS) services being added and the AEHF XDR and WGS Ka-band being unused in the "Command Ships" column. The Marine Corps is noted as utilizing AEHF XDR service for brigades ("Brigade" column) and WGS and UFO/GBS service for Corps-level communications ("Corps" column). Strike aircraft ("Strike A/C") similarly rely on AEHF XDR and UFO/UHF service.

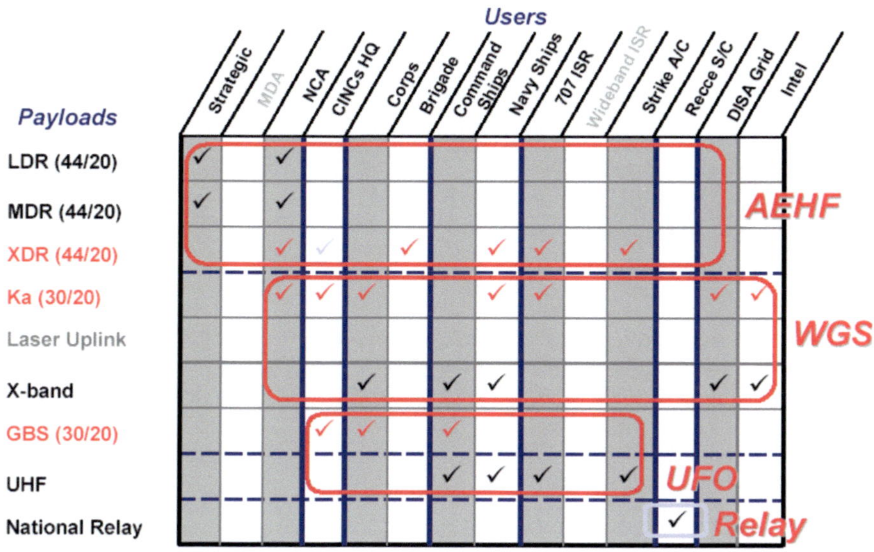

FIGURE D.1 Notional breakdown of Department of Defense satellite systems, by users. (A list of acronyms is provided in Appendix G.) SOURCE: Christine M. Anderson, Program Director, Military Satellite Communications Joint Program Office. 2002. "Transformational Communications," slide 6, presentation to Ground System Architectures Workshop, El Segundo, Calif., March 14.

WIDEBAND SERVICES

Wideband services currently are allocated into channelized communications channels provided by both commercial and DOD satellites. Current wideband services are provided by the Defense Satellite Communications System (DSCS) satellites, and by the GBS through an additional payload on the most recent UFO satellites. Services include unprotected and some secure data services, from 2.4 to 128 kb/s, voice channel capacity from 16 to 72 kb/s, and broadcast data and video services at up to 6 Mb/s. In addition to these services there are wideband space-to-space links provided by national space relays and commercial sources. The basic roadmap for DOD wideband services is shown in Figure D.2.

Current wideband satellite services provided by DSCS-Service Life Extension Program (SLEP) and GBS satellites will be augmented by the WGS, now under contract to the Boeing Company. WGS will provide two-way communications in X, K, and Ka bands, with broadcast services in K and X bands. The aggregate throughput of each system will be approximately 3.6 Gb/s, with a bus power capacity of 18 kW. According to Boeing, WGS will offer 4.875 GHz of instantaneously switchable bandwidth. The WGS system will provide capacity

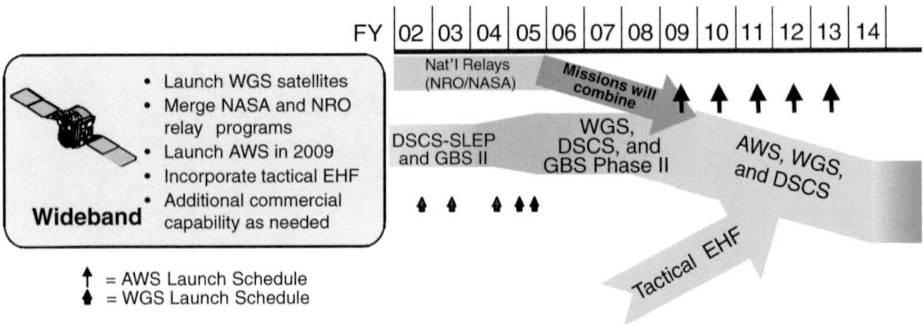

FIGURE D.2 Roadmap for Department of Defense wideband satellite services. (A list of acronyms is provided in Appendix G.) SOURCE: Adapted from Col J. Barry Patterson, Chief of Satellite Communications Division (J6S), U.S. Space Command, "Transformational Communications Study," February 28, 2002.

ranging from 1.2 Gb/s to more than 3.6 Gb/s to tactical users, depending on the mix of ground terminals, data rates, and modulation schemes employed. The WGS design includes 19 independent coverage areas that can be directed throughout the field of view of each satellite to serve warfighters between 65 degrees North and South latitudes. The connectivity capabilities of WGS enable users talk to other users with efficient use of satellite bandwidth, using subchannel routing techniques. A total of three WGS satellites have been contracted to date, with an initial launch scheduled for 2005.

The Wideband System (WBS) shown in the roadmap in Figure D.2 is being studied today under an overarching concept of the Transformational Communications Architecture (TCA). Over the past 2 years, this architecture has been developed to address the issues of (1) explosive growth of DOD bandwidth needs; (2) data efficiency, as circuit switching migrates to packet-based technologies; and (3) complexity of terminal needs to support current (legacy) systems. TCA will implement a gigabits-per-second Internet-like backbone in space, to provide worldwide high-speed packetized information and data services to mobile and fixed-site users. Its scope includes the space segment, providing radio frequency (RF) and laser communications linking air and space data between theater and continental United States users, a user terminal segment providing laser and RF user terminal development and production in concert with the space segment, and a terrestrial segment providing network control and interface services into the Global Information Grid (GIG).

Wideband Gapfiller System

The WGS program will provide the next generation of wideband communications for the DOD while maintaining compatibility with existing and pro-

TABLE D.1 General Characteristics of Wideband Gapfiller System

	Description
Primary function	High-capacity military communications satellite
Primary contractor	Boeing Satellite Systems
Satellite bus	Boeing 702
Weight	Approximately 13,000 lb at launch; 7,700 lb on-orbit
Orbit altitude	22,300 miles; geosynchronous
Payload	Transponded, crossbanded X- and Ka-band communications suite
Antennas	8 beams, transmit and receive X-band phased arrays; 10 Ka-band gimbaled dish antennas; 1 X-band Earth coverage
Capability	39 channels, 125 MHz each via digital channelized/router

SOURCE: Air Force Space and Missile Systems Center Web site <http://www.losangeles.af.mil/SMC/MC/wgs.htm>. Accessed May 5, 2004.

grammed X- and Ka-band terminals.[1] Broad WGS characteristics are provided in Table D.1. The WGS implementation plan calls for three to six geosynchronous spacecraft. The contract award was made in January 2001, and at the time of this writing, initial operational capability is planned for April 2006 and full operational capability for February 2007.[2] These new satellites will transmit an aggregate of several gigabits of data per second—up to 10 times the data flow of the satellites that the WGS will replace—though, as shown below, the objective maximum rate for the Navy's new WGS terminal supports, at most, 40 Mb/s for even the most advantaged Navy platforms.

The WGS constellation will supplement the two-way military X-band (7 to 8 GHz) communications capability now provided by the DSCS and the receive-only military Ka-band (20 to 21 GHz) downlink provided by the GBS. In addition, the WGS will provide a high-capacity two-way Ka-band capability to support mobile and tactical personnel. Early estimates indicated that one WGS satellite could provide transmission capacity up to 2.4 Gb/s. This capability alone exceeds the capacity of the entire existing DSCS and GBS constellations. Capacity gains will be matched by improved features, such as multiple high-gain spot beams that are particularly important for small terminal and mobile users.

Each WGS satellite supports 9 X-band beams and 10 Ka-band beams. Eight of the X-band beams are formed by separate transmitting and receiving phased-

[1]Information in this section was adapted from the Air Force Space and Missile Systems Center Web site on WGS <http://www.losangeles.af.mil/SMC/MC/wgs.htm>; SPAWAR PMW176, available at <http://enterprise.spawar.navy.mil/pmw176/products.htm>; and G. Elfers and S.B. Miller, 2002, *Future U.S. Military Satellite Communication Systems*, Aerospace Corporation, Los Angeles, Calif., available at <http://www.aero.org/publications/crosslink/winter2002/08.html>. All accessed May 5, 2004.

[2]DOD Press Release, "Department of Defense Releases Selected Acquisition Reports," August 15, 2003, available at <http://www.defenselink.mil/releases/2003/nr20030815-0374.html>. Accessed May 5, 2004.

array antennas that can adaptively form beams onto specific, desired coverage areas. The ninth X-band beam provides Earth coverage. The 10 Ka-band beams are formed by gimbaled dish antennas and include 3 beams with reversible polarization. An onboard digital channelizer divides the payload's overall communications capacity into 1,872 subchannels of 2.6 MHz each and independently switches each subchannel. The signals can be crossbanded from one frequency band to another, and any uplink coverage can be connected to any downlink coverage. Also, any uplink signal within one coverage area can be connected to any or all downlink coverages.

User terminals capable of operating within the several WGS frequency bands are a fundamental piece of the wideband architecture. The Air Force and Army are acquiring up to 200 lightweight, high-capacity quad-band ground multiband terminals (GMTs) for use with the WGS, DSCS, future Advanced Wideband System (AWS), and commercial satellite systems. In addition, the Army's Multiband/multimode Integrated Satellite Terminal (MIST) will provide up to a few megabits per second of capacity for on-the-move communications.

Navy Wideband Gapfiller Satellite Terminals

Within the Navy, the Super High Frequency (SHF)/Commercial Satellite Communications Division is responsible for procuring, fielding, and sustaining satellite communications (SATCOM) terminal systems to provide wideband communications services to the fleet. In addition to fielding SHF (X-band) satellite communications terminals (WSC-6) to the fleet, the division is inserting new technology to leverage the increase in capability in the SLEP and WGS programs.

Recent Navy documents indicate that the Navy is not planning to develop terminals to provide a full utilization of the SHF bandwidths. Rather, the Navy is focusing its future developments on terminals that are more modest in size and thus enable lower inherent bandwidth to the satellite. In particular, a recent article on the Navy's AN/WSC-6(V)XX SHF communications terminal program states that "four antenna aperture diameters are currently envisioned to be approximately 38, 60, 93 and 108 in. (3 ft, 5 ft, 7¼ ft, and 9 ft) although the actual sizes shall be proposed by offeror. . . . With its four antenna size variants, the AN/WSC-6(V)XX will see service on virtually all U.S. Navy and Military Sealift Command (MSC) surface ship classes. . . . The AN/WSC-6(V)XX Dual Antenna Handover System (DAHS) shall mitigate SATCOM disruptions caused by antenna handover switching when the AN/WSC-6(V)XX operates at full-duplex data rates from 64 kilobits per second to at least 24 megabits per second, *with an objective maximum of 40 megabits per second* (emphasis added)."[3] Thus, the

[3]See <http://www.fbodaily.com/archive/2002/10-October/11-Oct-2002/FBO-00185320.htm>. Accessed May 5, 2004.

Navy's long-term bandwidth capabilities, available through the WGS, will be limited to 40 Mb/s at most.

Marine Corps Wideband Gapfiller System Terminals

The lightweight multiband satellite terminal (LMST) is a tri-band SATCOM terminal capable of operation in the military X band and commercial C and Ku bands.[4] At present it is capable of military Ka-band "receive only," but there is a Program Objective Memorandum (POM) 04 initiative to improve the Ka band to two-way communications, synchronizing with the launch of the WGS. Each LMST can support four links and an aggregate data rate of 8.448 Mb/s and is upgradeable to 20 Mb/s. The LMST is a transit case system that can be moved by 2 high-mobility multipurpose wheeled vehicles (HMMWVs) and operated by a two-person crew.

The Marine Requirements Oversight Council directed that the LMST compete to retire the legacy suite of ground mobile forces SATCOM terminals to reduce the operational risk involved in maintaining the current 20-year-old terminals. The LMST approved acquisition objective of 50 systems has now been approved. The full approved acquisition objective replaces legacy ground mobile forces terminals, thus providing a ten-fold improvement in Marine Expeditionary Force bandwidth, plus reduced operations and maintenance costs on 20-year-old terminals and simplified support for just the SHF SATCOM terminal type.

Transformational Communications

The Transformational Communications (TC) program will provide a revolutionary leap forward in DOD space-based communications. It is part of an overall plan that calls for a new IPv6-based DOD network architecture, a new end-to-end cryptographic architecture based on IP security employing the High Assurance Internet Protocol Encryptor (HAIPE), a new Advanced Polar Satellite (APS) constellation, airborne laser communications terminals, and a new DOD network management architecture.

The TC implementation plan calls for a total of up to eight geosynchronous TC spacecraft and APS satellites connected by laser crosslinks. The entire TC program is still early in its development, with Phase B acquisition of the space segment starting in the fall of 2003 under two competing contractor teams; at the time of this writing, first launch is scheduled for December 2009. These new satellites will be capable of extremely high communications capacity.

Figure D.3 presents the current view of the TC system in context: TC is the shaded cloud in the center with interfaces to other systems as depicted. As can be

---

[4]LtCol Wayne R. Martin, USMC. 2003. *Health of Command Element Advocacy*, U.S. Marine Corps, Washington, D.C., February 10.

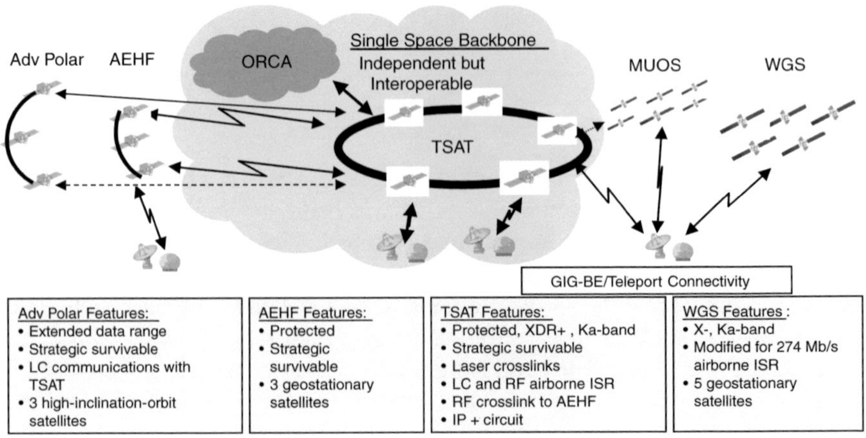

FIGURE D.3 Transformation communications system in context. (A list of acronyms is provided in Appendix G.) SOURCE: CDR J.J. Shaw, USN, "Transformational Communications Architecture," Office of the Chief of Naval Operations (N611), Washington, D.C., August 11, 2003.

seen, one of TC's objectives is to merge assured communications functions, over the long term, from the MILSTAR (Military Strategic, Tactical, and Relay) satellite and AEHF programs, and to utilize the AEHF system and Mobile User Objective System (MUOS) as edge systems for implementing information delivery to users, while providing the overall high-speed, space-space, and space-ground links.

Table D.2 presents the current view of the capabilities of an individual TC satellite. As can be seen, these satellites are currently expected to support a number of beams in a wide variety of bands ranging from 7 to 44 GHz. Each satellite will also perform onboard packet routing.

Terminals for Transformational Communications

The Navy and Marine Corps will need to develop new terminals to take proper advantage of TC. The Navy terminal program associated with TC has as an objective to develop an architecture, execute research and development (R&D), and prepare for the procurement of a single, multiband, multifunctional terminal suite. The suite would be configurable to meet diverse shipboard, submarine, air, and shore terminal requirements. Figure D.4 displays a notional terminal migration path to support TC links to Navy ships, and Table D.3 presents the Navy's baseline fielding schedule for TC Spiral 1 terminals. As shown, this baseline includes surface ships, submarines, and aircraft.

TABLE D.2 General Capabilities and Features of the Transformational Communications Satellite (TSAT)

Capability/Feature	Description
Notional TSAT Capabilities	
EHF Communications	44 GHz up; 20 GHz down
—Raw capacity per TSAT	0.8 to 3.1 Gb/s; does not include IP gain, link margin management, and so on (AEHF 0.2 to 0.3 Gb/s)
—Space-based IP router	Bandwidth on demand
—Waveform	XDR
—Input channels	40 active processed
—Output channels	17 active
—Nuller antennas	Two 80 in. 19 element EHF
—Multibeam antennas	One 40 in. EHF
—Gimbaled dish antenna	Six 24 in.
—Rx phased array	Two 10-beam
—Tx phased array	Two single-beam
Ka-band payload	30 GHz up; 20 GHz down
X-band payload	8 GHz up; 7 GHz down
Optical communications	Five laser heads
TSAT Payload Features	
Theater multibeam antenna	Two 80 in. 44/20 GHz, include single 8/7 GHz Tx/Rx
Gimbaled dish antenna	Six 24 in., include 44/20, 8/7, and 30 GHz
Active receive phased array	One (ten 44 GHz beams)
Active transmit phased array	Two (one 20 GHz beam each)
Waveform	Includes broadband communications XDR modes
Routing	Onboard packet
Transmitters	17 D/L, including one 20 GHz per ATPA, three 20 GHz and two 7 GHz for GDAs, eight 20 GHz and one 7 GHz for theater multibeam arrays
Apertures	5 optical, all single access (spiral 1) and 1 multiaccess (spiral 2)

NOTE: A list of acronyms is provided in Appendix G. SOURCE: MILSATCOM Joint Program Office, 2003, "Transformation Communications MILSATCOM Industry Day Brief," Air Force Space Command, El Segundo, Calif., March 5-7; and Michael Frankel, OSD, "Implementing the Global Information Grid (GIG)," presentation to the committee on June 27, 2003.

Navy's Active Role in Transformational Communications

Because the TC program will have an extremely important impact on naval communications, the Navy has been active in the development and management of the architecture. This role includes recent Navy leadership of the Transformational Communications Architecture (TCA) office.

It is essential for the Navy and Marine Corps to continue to engage as full partners in the ongoing design, refinement, and acquisition of the TC system. It is likely that numerous trade-offs will be made in the course of system design, and the Navy and Marine Corps must be knowledgeable and fully "in the loop" as

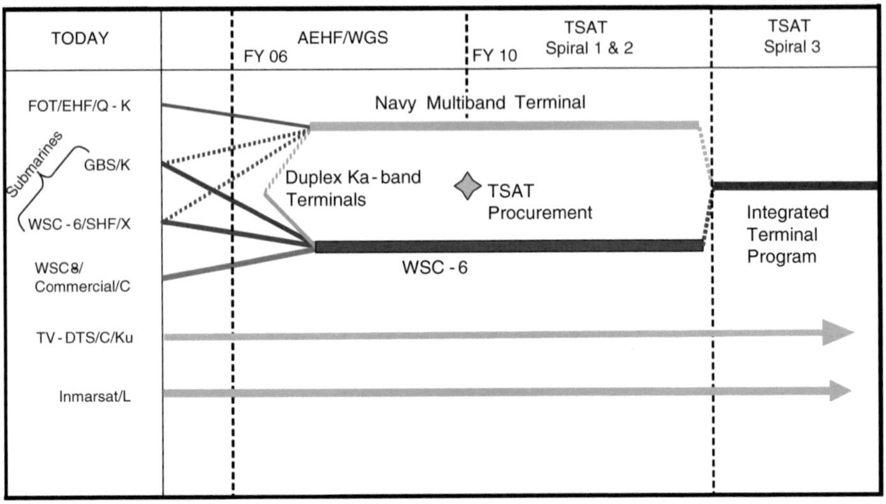

FIGURE D.4 Navy terminal migration strategy for Transformational Communications (TC): path to support TC links to Navy ships. (A list of acronyms is provided in Appendix G.) SOURCE: Michelle Bailey, PEO C4ISR, "Navy Transformational Communications (TC) Terminal Acquisition," May 20, 2003.

these decisions are made. This participation is not only the most effective way to influence implementation of Navy bandwidth needs, but it would also protect against the possibility that essential naval capabilities could be obscured or modified in the requirements versus cost trade-offs occurring during the course of program development.

To this end, there is concern that the Navy and Marine Corps have not developed an overall support strategy for TCA and other key space communications acquisitions. Such a strategy would need to have continuity over TCA's acquisition lifetime and should include seniority in key assignments, depth and knowledge of support staff, and communications regarding the status of acquisitions and key issues within naval stakeholding organizations.

While the TC architecture and acquisition program matures, it should be recognized that TC is an extraordinarily ambitious program, and it is fairly predictable that this program, like other ambitious space programs, will overrun its current cost projections and extend its currently projected launch dates. Thus, the Navy and Marine Corps will need to continue to provide upgrades to current capabilities as they migrate to the new communications capability, but they will need to ensure that robust operational capability be provided via existing commercial, UHF, MUOS, GBS, AEHF, and MILSTAR programs, first as the fallback path until TC dates are better understood, then as assured capability until the

TABLE D.3 Navy Baseline Transformational Communications Terminal Fielding Schedule

Funding Source	No. of Terminals Scheduled							
OPN Funded	FY09	FY10	FY11	FY12	FY13	FY14	FY15	Total
AGF		4						4
CG	4	2	4	4	5	6	2	27
CV/CVN/CVNX	4	2	4	4	4	4	2	24
DDG	5	4	5	8	10	12	9	53
LCC		2			2			4
LHA	2	2		2	2		2	10
LHD	2			4	4	4	2	16
LPD		1		6	5	3	2	17
LSD				5	5		2	12
LCS	1	1	1	1				4
SSBN/SSGN		4	4	4	6	2	2	22
SSN (Los Angeles Class)	2	3	4	11	11	9	9	49
SSN (Seawolf Class)			1	1	1			3
SSN (Virginia Class)			1	1	1	1		4
Operations (Shore)		6	7	20	10	3	3	49
Training (Ship/Submarine/Shore)	6							6
Test (Ship/Submarine/Shore)	5							5
Total OPN Funded	31	31	31	71	66	44	35	309
SCN Funded	FY09	FY10	FY11	FY12	FY13	FY14	FY15	Total
CV/CVN/CVNX				2		2		4
DDG					3	3	3	9
JCC(X)					2	2	2	6
LCS			4	4	4	1		13
Total SCN Funded			4	6	9	8	5	32
APN Funded	FY09	FY10	FY11	FY12	FY13	FY14	FY15	Total
E-6 Aircraft			5	5	5	5		20
UAV		2	7	10	12	11	9	51
UCAV						8	8	16
Total APN Funded	0	2	12	15	17	24	17	87

NOTES: OPN, other procurement, Navy; AGF, miscellaneous command ship; CG, guided missile cruiser; CV, aircraft carrier; CVN, nuclear-powered aircraft carrier; CVNX, future aircraft carrier; DDG, guided missile destroyer; LCC, amphibious command ship; LHA, amphibious assault ship (general purpose); LHD, amphibious assault ship (multipurpose); LPD, amphibious transport dock; LSD, landing ship, dock; LCS, landing craft support; SSBN, nuclear-powered ballistic missile submarine; SSGN, nuclear-powered guided-missile submarine; SSN, nuclear-powered attack submarine; SCN, ship and construction, Navy; JCC(X), joint command and control ship; APN, aircraft procurement, Navy; UAV, unmanned aerial vehicle; UCAV, uninhabited combat aerial vehicle.

SOURCE: Michelle Bailey, PEO C4ISR, "Navy Transformational Communications (TC) Terminal Acquisition," May 20, 2003.

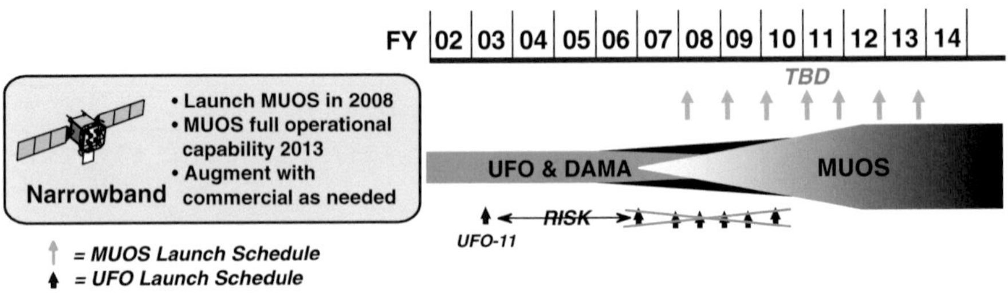

FIGURE D.5 Department of Defense narrowband satellite communications systems roadmap. (A list of acronyms is provided in Appendix G.) SOURCE: Adapted from Col J. Barry Patterson, Chief of Satellite Communications Division (J6S), U.S. Space Command, "Transformational Communications Study," February 28, 2002.

TC systems are mature and integrated, and then as backup until retirement of all legacy systems occurs.

NARROWBAND COMMUNICATIONS

Narrowband communications services are defined as 64 kb/s or less of raw channel capacity to a user device (platform). These services, primarily unprotected narrowband tactical circuit-switched communications, are now provided by eight UFO satellites, the first launched in 1993 and reaching design lifetime around 2008. The narrowband constellation now supplies two-way low-rate voice and tactical switched circuits to the fleet. Functions include command and control communications between the combatant commanders and their components; connectivity for command and control of tactical forces; connectivity for deployed Special Operating Forces; connectivity supporting rapid deployments of land, air, and naval forces worldwide; and connectivity for tactical communications in all operating environments.

Figure D.5 shows the nominal roadmap for the future development of narrowband (UHF) satellite communications systems. The UFO constellation is complete with the activation of the eleventh UFO satellite (UFO-11), which was successfully launched in December 2003. Because the design lifetime of the initial UFO satellites (15 years) is nearing an end, the Navy is leading the acquisition of the next-generation narrowband system, the Mobile User Objective System. This continues the 30-year history of naval leadership in narrowband space communications.[5]

[5]For a review of U.S. Navy involvement in narrowband communications, see Jerry Ingerski, SPAWAR, PMW 146, and Alfred Sapp, Naval Network Warfare Command, 2002, "Mobile Tactical Communications, The Role of UHF Satellite Constellation, and Its Successor, the Mobile User Objective System," paper and brief presented at the 2002 Military Communications Conference, Armed Forces Communications and Electronics Association (AFCEA) International, Anaheim, Calif., October 7-10.

The Mobile User Objective System is the successor to today's UHF Follow-On (UFO) system—UFO satellites were first launched in 1993 and UFO will remain the principal narrowband constellation until MUOS comes online. MUOS is currently expected to achieve initial operational capability in 2009, with full operational capability in 2013. In August 1996, the Deputy Undersecretary of Defense for Space was tasked by the Joint Space Management Board to further define the DOD Space Architect's Military Satellite Communications (MILSATCOM) recommended architecture and to develop an affordable transition roadmap for this system. In November 1996, the Navy volunteered to lead the Joint Mobile User Study, which analyzed three main areas: requirements, systems engineering, and costing and acquisition strategy, and from more than 100 mobile user narrowband requirements, the study determined the following eight primary requirements, in descending order of priority:

- Assured access,
- Netted communications,
- Communications on the move,
- Joint interoperability,
- Worldwide coverage,
- Point-to-point communications,
- Broadcast, and
- Polar coverage.

MUOS is being designed for compatibility with more than 50 types of existing UHF satellite communications terminals, including the AN/PSC-5 Spitfire, URC-133 Federated, ARC-210, and WSC-3. The Space and Naval Warfare Systems Command (SPAWAR) has succinctly described its rationale for the choice of the UHF band for user services as follows:

> There is a great need for the UHF portion of the spectrum because it gives the warfighter the ability to penetrate heavy weather, foliage, and concrete reinforced buildings. UHF generally includes the frequencies from 300 MHz to 3 GHz. The portion of the UHF spectrum that does the job for the warfighter today is from 200 MHz to 400 MHz. UFO satellites, the first launched in 1993, are the mainstay UHF communications for the mobile warfighter. They operate in the general range of 290-320 MHz Uplink, and 240-270 MHz Downlink. These frequencies are well suited for low-cost, low power, portable radios that reliably penetrate severe environments and offer assured access and netted communication.[6]

[6]CAPT James Loiselle, USN, Robert Tarleton, and Jerry Ingerski, Communications Satellite Program Office, Space and Naval Warfare Systems Command, San Diego, California. 1998. "The Next Generation Mobile User Objective System (MUOS)," American Institute of Aeronautics and Astronautics of Aeronautics and Astronautics, Inc., AIAA-98-5246.

This rationale is a sound basis for the development of the MUOS program.

Although the Under Secretary of the Air Force is the DOD Executive Agent for Space, the Navy has been delegated responsibility for unprotected, narrowband satellite development.[7] Accordingly, the Navy is procuring MUOS as the DOD narrowband satellite system for worldwide UHF communications. The Deputy Chief of Naval Operations for Warfare Requirements and Programs is the Navy's satellite communications program sponsor, and the Communications Satellite Program Office at SPAWAR is the Navy's Communications Satellite Acquisition Program Office. Two contractor teams are currently performing Phase II Concept Advanced Development, the updated operational requirements document is scheduled to be presented for Joint approval in early 2004, and acquisition will begin in mid-2004. As stated above, the MUOS constellation is planned for initial launch in 2009 and full operation in 2013.

As the Navy's only space program, MUOS is very important for the Navy. It is unclear how the Navy plans to integrate MUOS into the evolving GIG and TCA in order to meet launch dates and achieve initial operation prior to lifetime expiration of the current UFO constellation. As FORCEnet develops its communications requirements over the next year, it is important to clarify how MUOS will function as an edge system in the TCA. It may be possible to adjust the MUOS system design and capabilities during the upcoming advanced development phase to better accommodate Transformational Communications interfaces.

MUOS will be an unprotected, narrowband system supporting a worldwide, multiservice population of mobile and fixed-site users. MUOS space and ground segments will include a network of advanced UHF satellites and the ground infrastructure necessary to manage the information network, control the satellites, and interface with other systems of the GIG. The network management will include improved capability for dynamic bandwidth allocation so as to be more responsive to changing operational communications requirements. During the transition from the UFO constellation, the MUOS will serve a user population consisting of a mix of legacy and new terminals. The new terminals will be Joint Tactical Radio System (JTRS)-compliant, designed to provide the mobile user with higher data rates and an improved link margin.

It is clear that MUOS, as a narrowband system, will not resolve Navy and Marine Corps current and future bandwidth needs. For the purposes of MUOS, narrowband is defined to be 64 kb/s or less of raw channel capacity to a user

---

[7] Information in this section is derived from the MUOS Draft Solicitation, dated September 10, 2003; Robert Tarleton, Deputy Program Manager, Space and Naval Warfare Systems Command, 2003, "Mobile User Objective System (MUOS) Brief to the Naval Studies Board," presented on November 14, 2003; and CAPT James Loiselle, USN, Robert Tarleton, and Jerry Ingerski, Communications Satellite Program Office, Space and Naval Warfare Systems Command, San Diego, California, 1998, "The Next Generation Mobile User Objective System (MUOS)," American Institute of Aeronautics and Astronautics of Aeronautics and Astronautics, Inc., AIAA-98-5246.

device (platform). In many cases, the planned user date rates will probably be significantly slower than this upper bound, on the order of 2.4 to 9.6 kb/s for handheld devices in urban terrain and 32 kb/s for large aircraft or submarines. However, MUOS does provide high-availability connectivity for basic communications needs for the mobile tactical user, who now relies on the oversubscribed UFO system.

In addition to developing a better understanding of the interface between the TCA and the MUOS system, there are two areas in which the MUOS program may be able to adjust to become a more effective system. First, because the MUOS system is backward-compatible with the UFO system, it specifies a set of narrowband channels that can be allocated to users (5 or 25 kHz wide). Fixed channelization is an inefficient use of bandwidth, particularly for the burst data communications likely to dominate MUOS traffic. During terminal development, the Navy may need to consider baselining the capability to multiplex multiple narrowband channels to form broader channels (100 kHz or more) and to allow utilization of the aggregate as a single channel of higher bandwidth. Second, because MUOS is an unprotected communications asset, it should not be used as the sole means of communications with any tactical user whose combat capabilities are reliant upon satellite communications. The MUOS system may be able to accommodate increased robustness against jamming through error protection or increased nulling capability in order to reduce this limitation.

ASSURED SECURE COMMUNICATIONS SERVICES

Assured communications capability (supplied by the MILSTAR and AEHF satellites) is the most recently developed of the families of military communications available from space. The objective of the assured communications capability is to provide worldwide, secure, survivable, protected communications for U.S. and allied forces. The capabilities are now supplied by the series of MILSTAR geosynchronous communications satellites and will be augmented by the AEHF satellites now being readied for an initial launch in 2006.

In the longer term, the TCA program will provide higher-bandwidth assured capabilities. These series of satellites are hardened in multiple ways to ensure the survivability of communications in case of a spectrum event of natural or hostile force actions. In fact, the MILSTAR system is the only protected and truly survivable space communications system available for U.S. military forces. It provides the unique capability for communications that the U.S. executive branch and DOD top commanders have to maintain positive command and control of strategic forces in the event of concerted and sophisticated enemy action. It is essential that the Navy communications strategy incorporate an understanding of the protected nature of the assured communications assets and plan capacities to retain essential warfighting capability in the event of loss of most or all of its unprotected communications links. The roadmap for assured space communica-

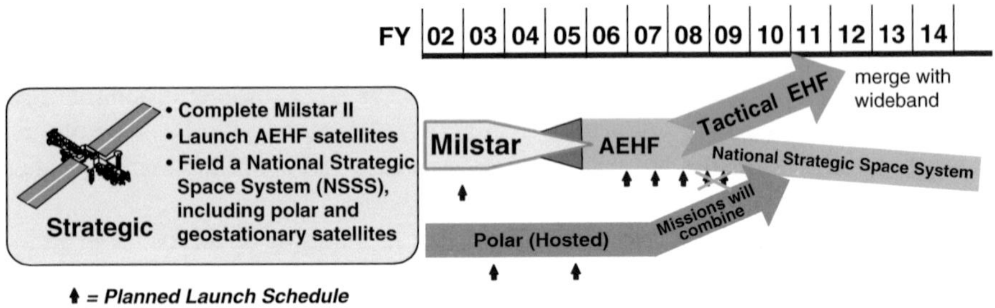

FIGURE D.6 Department of Defense assured space communications roadmap. (A list of acronyms is provided in Appendix G.) SOURCE: Adapted from Col J. Barry Patterson, Chief of Satellite Communications Division (J6S), U.S. Space Command, "Transformational Communications Study," February 28, 2002.

tions capabilities is shown in Figure D.6. These programs and their capabilities are discussed below.

MILSTAR

With the launch of the last MILSTAR II communications satellite in the spring of 2003, the Air Force completed implementing the MILSTAR constellation of five satellites. The early MILSTAR satellites provided assured communications link capacities of a few tens of kb/s, and MILSTAR II provides an upgraded capacity of up to 1 Mb/s. The MILSTAR constellation provides secure transmission of critical command and control information (including voice, data, and imagery, as well as voice and video teleconferencing) between deployed forces and the command structure. MILSTAR provides rapid link establishment and switching, allowing space-space transmission of data and thus eliminating many space-to-ground communications hops.

The capabilities of MILSTAR, and in particular MILSTAR II, were quickly pressed into service in support of Operation Iraqi Freedom. Early lessons learned from the war are indicated in the value of increased bandwidth to warfighting capability. For example, one review stated, "The real star in supporting/exploiting ISR was MILSTAR II: high data rate, antijam, encrypted and 100% available transmission of NRO/CIA processed intelligence to theater, remote tasking of Global Hawk, critical communications support of special operations, and rapid retargeting of Tomahawk missiles."[8]

[8]Loren B. Thompson. 2003. "ISR Lessons of Iraq," presentation at Defense News ISR Integration Conference, November 18.

U.S. Navy ships and tactical users routinely use MILSTAR and will, in the future, utilize the AEHF systems as an integral part of their communications structure. These systems provide the assured communications required to ensure continuity of command in all threat environments.

Advanced Extremely High Frequency Satellite System

The Advanced Extremely High Frequency (AEHF) program now in development will provide a series of advanced, protected communications satellites that will augment and eventually replace the current MILSTAR system as it reaches its design lifetime.[9] The overall objective of the AEHF program is to develop and field a constellation of four geosynchronous AEHF satellites to provide worldwide secure, survivable, protected communications that are backward-compatible with MILSTAR but that significantly advance the capacity and capabilities for assured protected worldwide communications.

The AEHF system will improve ease of operations, facilitate satellite control and monitoring, and effectively interface with evolving terminal designs. AEHF will consist of four cross-linked satellites covering the globe from 65 degrees North to 65 degrees South latitude, providing 10 times the data rate available through MILSTAR. The system provides uplinks in the EHF spectrum and downlinks in SHF spectrum.

The AEHF system is currently in its system development and demonstration/production acquisition phase, with the first launch currently scheduled for 2006 and initial operational capability scheduled for 2007. The system will provide backward compatibility with the existing MILSTAR low data rate (LDR) and medium data rate (MDR) services. It will also provide a new, expanded data rate (XDR) service. It will support a range of user data rates between 75 b/s and 8 Mb/s. Table D.4 presents the general characteristics of the AEHF system.

Advanced Extremely High Frequency Aircraft Terminals

Naval aircraft may employ the family of advanced beyond-line-of-sight terminals (FAB-T) that are currently being procured by the Air Force. FAB-T will develop robust, secure, survivable EHF voice and data satellite communications terminals for nuclear and conventional forces. FAB-T variants will provide ground and airborne command posts and other aircraft with connectivity to MILSTAR and AEHF satellites, while providing an open architecture terminal to support future increments for WGS, EHF payloads on polar and UFO satellites, GBS payloads, and TSAT.

---

[9]Information in this section is derived from the Air Force Space and Missile Command Web site at <http://www.losangeles.af.mil/SMC/MC/aehf.htm>. Accessed May 5, 2004.

TABLE D.4 General Characteristics of the Advanced Extremely High Frequency Satellite System

	Description
Primary function	Worldwide, secure, survivable satellite communications
Primary contractor	Lockheed Martin Satellite Systems
Satellite Bus	A2100 line
Weight	Approximately 13,100 lb at launch; 9,000 lb on-orbit
Orbit altitude	22,300 miles; Geostationary
Payload	Onboard signal processing, crossbanded EHF/SHF communications
Antennas	8 gimbaled dishes, 1 EHF and 2 SHF phased arrays, 2 Earth coverage horns, 2 crosslinks
Capability	Data rates from 75 b/s to approximately 8 Mb/s

NOTE: A list of acronyms is provided in Appendix G.
SOURCE: Accessed from <http://www.losangeles.af.mil/SMC/MC/docs/aehf_bw_200.pdf.>. Accessed May 5, 2004.

Advanced EHF Surface Ship, Shore, and Submarine Terminals

The Navy Multiband Terminal is planned to support AEHF. In November 2003, SPAWAR awarded competitive 30-month contracts for four prototype terminals under direction of PMW 176-3. The acquisition phase is currently expected to begin in 2006 and to produce as many as 300 terminals.

COMMERCIAL SATELLITE SERVICES

In addition to DOD-dedicated space communications assets, the Navy routinely augments bandwidth by acquiring large amounts of communications services, including Intelsat, Inmarsat, Iridium, and the Defense Satellite Transmission Service (DSTS). In particular, the Navy is a major user of Inmarsat and Iridium for narrowband and voice services and Intelsat and DSTS-Global (DSTS-G) for wideband transponder-base services, and it is considering new ventures such as Boeing's proposed Connexion service. These services are of increasing use as a quick way to acquire large amounts of bandwidth, but they are unprotected and require, in most cases, yet another set of dedicated terminals and connections into the local network.

ADVANCED DIGITAL NETWORK SYSTEM

For platforms with multiple satellite links, it is possible in principle to route traffic dynamically over links, depending on traffic type or priority, current operational or system status, and so forth. This capability is currently provided by the Advanced Digital Network System (ADNS). In fact, the Navy has put much of this capability into its network routers and currently plans to further extend

ADNS functionality via procurement during 2004. This upgraded ADNS capability will provide a useful new feature, namely, truly dynamic routing of traffic over multiple satellite links, and perhaps even over other forms of RF links, rather than routing data according to present-day preconfigured rules.

Looking farther into the future, the current ADNS capability needs further evolution from its current role of carrying Red (secret-level) traffic to a new Black ADNS system that is compatible with the GIG IP transport architecture. The details of this evolution are beyond the scope of this report, but the Navy will need to follow Black ADNS development as the Navy's communications capabilities evolve and become compatible with the GIG architecture.

HYBRID NETWORK ARCHITECTURES

Satellite communications links have traditionally functioned as stand-alone communications pipes that directly tie a ship, submarine, or ground unit to its shore base. Going forward, naval units will be linked into a mesh network that also includes direct peer-to-peer data links (e.g., ship to ship); multihop wireless networks with ship, air, and land components; and high-capacity directional radio links, such as the Common Data Link.

This new networking approach offers enormous promise for robust, high-capacity communications for naval forces. In particular, space communications can play a greatly increased role in such hybrid systems because they do not need to provide a complete solution—the inherent weaknesses of satellite communications (e.g., high latency, weather issues with optical links, and so on) can be compensated by augmentation with other communications links as contributory.

In short, the great strength of satellite communications—ubiquitous, worldwide communications from anywhere to anywhere—can be married to robust, high-capacity local communications to form an overall network that is far more capable than any of its components in isolation. Figure D.7 provides one schematic illustration of such a hybrid network. The figure shows a naval force linked at very high bandwidth, and with great resilience, into the GIG via a hybrid optical/RF network. Optical links are employed between satellites and high-altitude aircraft (such as Global Hawk or E2-C). In general, these links will operate above the atmospheric effects that make direct optical links to the sea surface or ground problematic. The aircraft in turn may relay traffic down to ships, submarines, or Marine Corps forces via RF communications such as the Man Portable Common Data Link that can provide high capacity (500 Mb/s or higher) with jam-resistant links. As a result, even relatively small surface platforms could achieve total bandwidths of a gigabit per second or more in all weather conditions, plus high degrees of assurance and availability in the face of adversary jamming.

Other hybrid network architectures have equally appealing properties. For example, a large-deck ship may contain high-capacity satellite links and then in

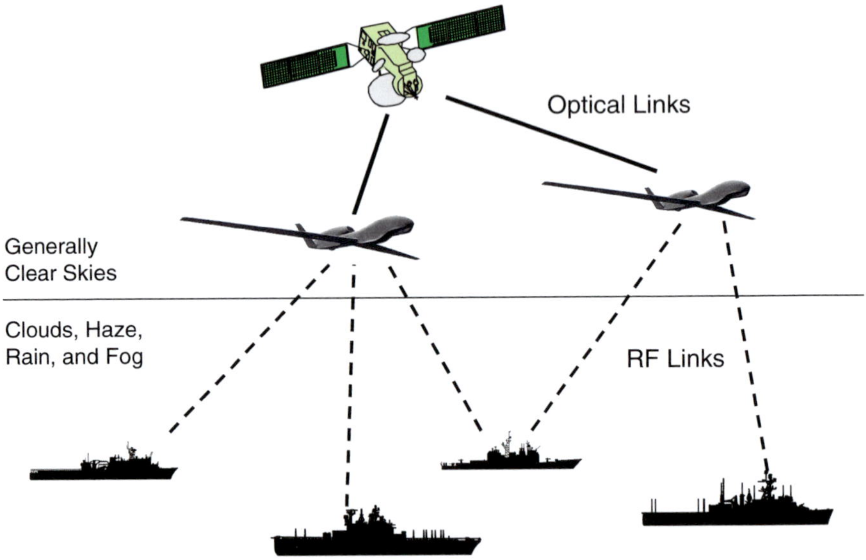

FIGURE D.7 Notional high-bandwidth hybrid network with optical and radio-frequency (RF) links.

turn act as a communications hub for smaller ships in its neighborhood. These ships may be linked by near-Earth RF links running as a peer-to-peer wireless network. For example, the JTRS Wideband Network Waveform may be used in this way. Elevated assets may be included in the mix in order to achieve beyond-line-of-sight connectivity from the large-deck hub to smaller platforms. In this approach, only the large-deck ships require their own large satellite aperture; smaller platforms then achieve high-bandwidth communications with the relatively small, nonstabilized antennas used by wireless ad hoc networks.

In summary, it would be a great mistake to analyze or acquire space communications capability in isolation from other naval communications capabilities. Hybrid space/near-Earth communications networks are likely to provide much higher performance, and much greater robustness, than an all-satellite approach, and may become a necessary component of future naval communications systems.

E

Biographies of Committee Members and Staff

Antonio L. Elias (Co-Chair) is executive vice president and general manager for advanced programs at Orbital Sciences Corporation. He had served as Orbital's chief technical officer from 1996 to 1997, as the company's corporate senior vice president from 1992 to 1996, and as its first vice president for engineering from 1989 to 1992. From 1987 to 1991, he led the technical team that designed and built the Pegasus air-launched booster, flying as a launch vehicle operator on the carrier aircraft for the rocket's first and fourth flights. He also led the design teams of Orbital's Advanced Photovoltaic and Electronic Experiments (APEX) and SeaStar satellites and X-34 hypersonic research vehicle. Dr. Elias went to Orbital from the Massachusetts Institute of Technology (MIT), where he held various teaching and research positions, including the Boeing Chair in the Department of Aeronautics and Astronautics. He is a member of the National Academy of Engineering and a fellow of the American Institute of Aeronautics and Astronautics (AIAA). His awards include the AIAA Engineer of the Year for 1991, the AIAA Aircraft Design Award, and the American Astronautical Society Brouwer Award. He is also a corecipient of the National Medal of Technology and the National Air and Space Museum Trophy. Dr. Elias is a member of the National Research Council's (NRC's) Naval Studies Board.

William D. Smith (Admiral, USN, Ret.) (Co-Chair) retired in 1993 after 38 years of active duty service. At present Admiral Smith is a senior fellow at the National Defense University with the Chairman of the Joint Chiefs of Staff's Capstone Program. His background is in Navy planning, programming, budgeting, and operational issues, principally within the submarine force. His last assignment was as U.S. military representative to the North Atlantic Treaty Organization's Military Committee in Brussels, Belgium. In addition, he served in a number of high-ranking capacities for the Chief of Naval Operations, such as Deputy Chief of Naval Operations for Logistics and Navy Program Planning from 1987 to 1991 and as director, Fiscal Management Division/Comptroller of the Navy from 1985 to 1987. His decorations include the Defense Distinguished Service Medal, the Distinguished Service Medal with Gold Star, the Legion of Merit with Three Gold Stars, the Meritorious Service Medal with Gold Star, and the Navy Commendation Medal. Admiral Smith has served on numerous scientific boards and advisory committees, including as chair of the recent Naval Space Panel Review for the Undersecretary of the Navy. He is a member of the NRC's Naval Studies Board.

Alan Berman is a part-time employee at the Applied Research Laboratory of Pennsylvania State University (ARL/PSU) and at the Center for Naval Analyses (CNA). At ARL/PSU, Dr. Berman provides general management support and program appraisal. At CNA, he assists with analyses of Navy research and development investments, space operations capabilities, information operations, and command, control, communications, computers, intelligence, surveillance, and reconnaissance (C4ISR) programs. His previous positions include serving as dean of the Rosenstiel School of Marine and Atmospheric Sciences at the University of Miami, where he was responsible for the graduate programs in physical oceanography, marine biology, geology, geophysics, applied ocean science, and underwater acoustics; and as director of research at the Naval Research Laboratory, where he administered broad programs in basic and applied research. Dr. Berman has served on numerous scientific boards and advisory committees, including as a member of the Free Electron Laser Oversight Board that advises the Department of Energy's Jefferson National Laboratory.

E. Ann Berman is founder and president of Tri-Space, Inc., a remote-sensing and software engineering company serving a broad range of environmental and security areas. Her research interests include remote sensing, hydrogeologic modeling, geographic information systems development, and the development of software for environmental management and surveillance. (Her remote-sensing work covers the spectral range from visible through thermal infrared, but includes working knowledge of the radio-frequency spectrum.) From 1984 to 1988, Dr. Berman served as the Deputy Assistant Secretary of the Navy for Command, Control, Communications, Intelligence, and Space. She has served on numerous scientific boards and advisory committees, including the NRC Committee on Environmental Information for Naval Use. Dr. Berman is a member of the recent Naval Space Panel Review for the Undersecretary of the Navy.

Thomas C. Betterton (Rear Admiral, USN, Ret.) is a visiting professor and space technology chair at the Naval Postgraduate School. Admiral Betterton retired after 35 years in the U.S. Navy, having served as a naval aviator and aerospace engineering duty officer, in addition to duties at the National Reconnaissance Office. He holds a B.S. in electrical engineering from the University of

Notre Dame and S.M. and E.A.A. degrees in aeronautics and astronautics from MIT. He has served on numerous scientific and advisory committees for the Department of Defense (DOD) and the National Aeronautics and Space Administration (NASA), including as vice chair of the International Space Station Management and Cost Evaluation Task Force. Admiral Betterton is a fellow of the AIAA.

Charles F. Bolden, Jr. (Major General, USMC, Ret.) is an independent consultant and former senior vice president at TechTrans International, Inc. He retired from the U.S. Marine Corps after 34 years of service. As a naval aviator, General Bolden flew more than 100 missions into North and South Vietnam, Laos, and Cambodia. In 1981, he became an astronaut and later flew the space shuttle on four flights. He was appointed assistant deputy administrator for NASA from 1992 to 1993 and subsequently served for a year as Deputy Commandant of the U.S. Naval Academy. General Bolden's command positions include that of Deputy Commanding General, First Marine Expeditionary Force (I MEF); Commanding General, I MEF (Forward) in support of Operation Desert Thunder in Kuwait; Deputy Commander, U.S. Forces, Japan; and his final tour as Commanding General, 3rd Marine Aircraft Wing.

John F. Egan is an independent consultant, having retired in 1998 as vice president for corporate development at Lockheed Martin Corporation. During his tenure at Lockheed, Dr. Egan was responsible for providing support to three successive chief executives in defining and implementing strategic plans to consolidate the defense industry. These included the merger of the Lockheed Corporation with the Martin Marietta Corporation and the acquisition of the defense segment of the Loral Corporation. During these mergers, Dr. Egan provided leadership throughout the entire transaction cycle involving industry and market analysis, deal negotiations, antitrust filings, and transition planning and execution. He has a broad understanding of Navy programs, business and strategic planning, and acquisition and policy. An electrical engineer by training, Dr. Egan is a former chief scientist for the Chief of Naval Operations and has extensive experience with electronic and information warfare. He has served on numerous scientific boards and advisory committees, including as a member of the Chief of Naval Operations Executive Panel.

Brig "Chip" Elliott is principal engineer at BBN Technologies, where he has led the design and successful implementation of a number of secure, mission-critical networks based on novel Internet technology for the United States, Canada, and the United Kingdom (NTDR, Iris, Bowman) and has acted as senior adviser on a number of national and commercial networks, including three low-Earth-orbit satellite constellations (Discoverer II, Space-Based Infrared Systems-Low, and Celestri/Teledesic) and Boeing's Connexion system. Mr. Elliott has particular expertise in wireless Internet technology, mobile ad hoc networks, quality-of-service issues, and novel routing techniques. He is currently leading the design and build-out of a very highly secure network protected by quantum cryptography. He holds some 70 patents pending or issued on network technology and has served on numerous scientific boards and advisory committees such as the Army Science Board and Defense Science Board. Mr. Elliott is a member of the NRC's Naval Studies Board.

Richard Fleeter founded AeroAstro, Inc., in 1988 and has been its chief executive officer and a member of the board of directors since the company's inception. A leading proponent of spacecraft miniaturization, he has led the development of several miniature satellites and subsystems. Over the past nearly 15 years, he has authored books on ways to reduce costs in space so significantly that many applications never before even considered are now practical. As an example, NASA can now fly microgravity missions in space for the price of a sounding rocket payload, and companies can monitor their remote assets via the World Wide Web for the cost of a set of Global Positioning System receivers. Prior to founding AeroAstro, Dr. Fleeter was a senior scientist at the California Institute of Technology's Jet Propulsion Laboratory and a project engineer at TRW, where he received a commendation for his contribution to the successful rescue of the communications satellite Tracking and Data Relay Satellite-1.

Lee M. Hammarstrom is special assistant for space and information technology to the director at the Applied Research Laboratory/Pennsylvania State University (ARL/PSU). Previously, Mr. Hammarstrom was the first chief scientist at the National Reconnaissance Office (NRO) and chief scientist at the Office of the Secretary of Defense for Command, Control, Communications, and Intelligence. In addition, he held various positions at the Naval Research Laboratory in remote sensing, reconnaissance, and intelligence leading to the creation of the Space System Engineering Division. Mr. Hammarstrom was named NRO Pioneer in 2002 for his 40 years of contributions to national reconnaissance.

Donald G. Hard (Major General, USAF, Ret.) is an independent consultant and sole proprietor of Hard Enterprise. General Hard's 33-year Air Force career focused on the development of space systems, culminating in his service as director of Space and Strategic Defense Initiative Programs, Office of the Secretary of the Air Force for Acquisition. In this position he provided program management direction for the development and procurement of Air Force surveillance, communications, navigation and weather satellites, space launch vehicles, antisatellite weapons, and ground-based and airborne strategic radars, communications, and command centers. After retiring from the Air Force in 1993, General Hard held several senior positions in defense-related companies, including Aerospace Corporation, Logicon/Northrop Grumman, and bd Systems, Inc. General Hard has served on numerous advisory committees related to the civilian and military development of space, including the NRC Committee on Space Facilities.

Robert E. Lindberg is president and executive director of the National Institute of Aerospace at Langley Research Center. (This institute for research and graduate education was created to carry out cutting-edge aerospace and atmospheric research, develop new technologies for the nation, and help inspire the next generation of scientists and engineers.) Dr. Lindberg's research interests include the development of rockets and satellites and the conceptual design of experimental spacecraft. Prior to his current position, he was employed at Orbital Sciences Corporation, where he led the industry/government team that developed the X-34 reusable launch vehicle testbed, led the development of the APEX satellite for the Air Force, and earlier contributed to the design of the Pegasus launch vehicle. His earlier career positions include service as research scientist and branch head in the Naval Center for Space Technology at the Naval Research Laboratory. Dr. Lindberg is a fellow and current president of the American Astronautical Society.

George O. Nossaman is director of space communications and electronics at BAE Systems, where he manages programs that develop and deliver radiation-hardened technology, computers, and subsystems for U.S. space programs. Prior to his current position, he served as director for technical operations and strategic planning at Lockheed Martin Federal Systems, where he directed technology in submarine combat systems, acoustic surveillance systems, space systems, and radiation-hardened electronics and space components. Earlier in his career, he served as senior program manager at IBM-Federal Systems, where he was responsible for their NASA-Johnson Space Station Control Center, the Space Shuttle Mission Control Center, and the Space Shuttle Mission Simulator programs.

C. Kumar N. Patel is chairman of the board of Pranalytica, Inc., and professor of physics and former vice chancellor of research at the University of California at Los Angeles. Until 1993, Dr. Patel served as executive director of the Research, Materials Science, Engineering, and Academic Affairs Division at AT&T Bell Laboratories. A member of the National Academy of Engineering and the National Academy of Sciences, he has an extensive background in several fields, including materials, lasers, and electro-optical devices. During his career at AT&T, which began in 1961, he made numerous seminal contributions in several fields, including gas lasers, nonlinear optics, molecular spectroscopy, pollution detection, and laser surgery. Dr. Patel has served on many government and scientific advisory boards. He is past president of Sigma Xi and the American Physical Society. He has received numerous honors, including the National Medal of Science, for his invention of the carbon dioxide laser.

Gene H. Porter is an independent consultant in matters relating to national security planning and weapons systems development. His current clients include the Center for Naval Analyses and the Institute of Defense Analyses, for which he works on research and development matters. Most recently, Mr. Porter has been supporting the Office of the Secretary of Defense in defining the detailed defense planning scenarios that are intended to guide the development of U.S. military force posture and modernization programs through the end of the decade. This analytic work has involved an all-source examination of potential threats, including space-based threats, and potential U.S. responses to them. Before assuming this work, Mr. Porter served as director of Acquisition Policy and Program Integration for the Office of the Undersecretary of Defense for Acquisition, where he was responsible for long-range planning, programming, and budgeting matters on new military warfare systems. His earlier career included various staff and line management positions at Sanders Corporation in the development and manufacture of military and commercial electronics systems, including mine and undersea warfare systems. Mr. Porter has served on numerous scientific and advisory committees and was chair of the NRC Committee for Mine Warfare Assessment.

Joseph B. Reagan is retired vice president and general manager of research and development at Lockheed Martin Missile and Space and retired vice president and corporate officer of the Lockheed Martin Corporation. Dr. Reagan joined Lockheed as a scientist in 1959, where he led the Space Instrumentation Group for 10 years and was responsible for the development and on-orbit deployment of more than 20 scientific payloads for NASA and the DOD. His research interests included the areas of space sensors, radiation belt and solar particles, nuclear weapons effects, and the effects of radiation particles on spacecraft systems. Later, as general manager of the Research and Development Division at Lockheed, he led more than 750 scientists and engineers in the development of advanced technologies in the fields of optics, electro-optics, information software, cryogenics, guidance and controls, electronics, and materials. Today Dr. Reagan is a director of Southwall Technologies, Inc., a company that manufactures energy-selective thin films for the automotive, electronics, and architectural industries. Dr. Reagan is a member of the National Academy of Engineering, a fellow of the AIAA, and vice chair of the NRC's Naval Studies Board.

Dwight C. Streit is vice president of science and technology for Northrop Grumman Space Technology. He is responsible for the development of the foundation technology required for advanced space systems, including high-performance electronics, antennas, propulsion, and structures. His prior research interests include semiconductor materials, devices and circuits for applications up to 220 gigahertz, infrared sensors, optical communications systems, and phased-array radar products. He is a fellow of the Institute of Electrical and Electronics Engineers and a member of the NAE.

H. Gregory Tornatore recently retired as special assistant to the department head for business development information and manager of special security at the Applied Physics Laboratory/Johns Hopkins University (APL/JHU). Previously he had served in a number of managerial positions, including that of program area manager for Defense Communications Systems. His research interests include strategic and tactical command, control, and communications (C3); over-the-horizon targeting systems; wide-area surveillance and reconnaissance; C3 systems vulnerability assessment; test and evaluation of major communications systems; satellite communications systems and architectures; command and control information processing; information operations; and communications networks. Mr. Tornatore has served on numerous scientific boards and advisory committees, including the NRC Committee to Review Department of Defense C4I Plans and Programs.

David A. Whelan is vice president and general manager of Boeing's Phantom Works Division. Before joining Boeing in 2001, Dr. Whelan was director of the Tactical Technology Office at the Defense Advanced Research Projects Agency (DARPA), where he led the development of enabling technologies, such as unmanned vehicles and space-based moving target indicator radar systems. Before serving with DARPA, Dr. Whelan held several positions of increasing responsibility with Hughes Aircraft. His high-technology development experience also includes roles as a research physicist for Lawrence Livermore National Laboratory, as well as being one of four lead engineers assigned to the design and development of the B-2 Stealth Bomber Program at Northrop Grumman. Dr. Whelan has served on numerous scientific boards and advisory committees, including the NRC Committee on Autonomous Vehicles in Support of Naval Operations.

Dell P. Williams III is an independent consultant in space systems and satellite communications. He recently retired as senior technical advisor to the president and chief executive officer of Teledesic Corporation, where he was responsible for technical oversight of the development of the Teledesic Network, involving a constellation of nongeostationary satellites providing worldwide access to broadband telecommunications services. Mr. Williams's background is in commercial satellite communications, space systems, space technology, and information assurance. Before joining Teledesic, he served as vice president of Deskin Research Group for Systems Engineering; vice president for Electronic Defense Systems at ARGOSystems, a wholly owned Boeing subsidiary; director of Advanced Programs at Lockheed Martin Missiles and Space Company; and director of Space Systems at NASA Headquarters.

Staff

Charles F. Draper is acting director at the National Research Council's Naval Studies Board. He joined the National Research Council in 1997 as program officer, then senior program officer, with the Naval Studies Board and in 2003 became associate director. During his tenure with the Naval Studies Board, Dr. Draper has served as the responsible staff officer on a wide range of topics aimed at helping the Department of the Navy with its scientific, technical, and strategic planning. His recent efforts include topics on network-centric operations, theater missile defense, mine warfare, and nonlethal weapons. Prior to joining the Naval Studies Board, he was the lead mechanical engineer at Sensytech, Inc. (formerly S.T. Research Corporation), where he provided technical and program management support for satellite Earth station and small-satellite design. He received his Ph.D. in mechanical engineering from Vanderbilt University in 1995; his doctoral research was conducted at the Naval Research Laboratory (NRL), where he used an atomic force microscope to measure the nanomechanical properties of thin-film materials. In parallel with his graduate student duties, Dr. Draper was a mechanical engineer with Geo-Centers, Inc., working on-site at NRL on the development of an underwater x-ray backscattering tomography system used for the nondestructive evaluation of U.S. Navy sonar domes on surface ships.

Michael L. Wilson was a program officer at the Naval Studies Board of the National Research Council from 2002 to August 27, 2004. From 1998 to 2002, Dr. Wilson was an assistant professor of physics at the University of Tulsa, where his research focused on granular dynamics under microgravity. From 1996 to 1998, Dr. Wilson was a visiting assistant professor of physics at Clemson University, where he helped establish a laboratory to study novel thermoelectric materials. Prior to working at Clemson, Dr. Wilson was a National Research Council associate at the Naval Research Laboratory, where he worked on ceramic magnetic materials synthesis and characterization as well as studies of superconductivity in thin metal films. He holds a Ph.D. in physics from Michigan State University and a B.A. in physics from Grinnell College.

F

Agendas for Committee Meetings

JUNE 26-27, 2003
KECK CENTER OF THE NATIONAL ACADEMIES
WASHINGTON, D.C.

Thursday, June 26, 2003

Closed Session: Committee Members and NRC Staff Only

0830 CONVENE—Welcome, Opening Remarks, Introductions
—Dr. Antonio L. Elias, Committee Co-Chair
—ADM William D. Smith, USN (Ret.), Committee Co-Chair
—Dr. Charles F. Draper, Associate Director, Naval Studies Board (NSB)

0900 COMPOSITION AND BALANCE DISCUSSION
—Dr. Dennis Chamot, Associate Executive Director, Division on Engineering and Physical Sciences, National Research Council

1000 COMMITTEE DISCUSSION—Terms of Reference, Study Plans, Other Issues
—Dr. Antonio L. Elias, Committee Co-Chair
—ADM William D. Smith, USN (Ret.), Committee Co-Chair
—Dr. Charles F. Draper, Associate Director, NSB

Data-Gathering Meeting Not Open to the Public: Classified Discussion (Secret)

1200 SPACE AND NAVAL WARFARE SYSTEMS COMMAND SPACE FIELD ACTIVITY—Naval Space Systems and Future Developments to Meet Operational Requirements
—RADM Rand H. Fisher, USN, Commander, Space and Naval Warfare Systems Command (SPAWAR) Space Field Activity; Director, Naval Space Technology Programs

1400 NAVY TECHNICAL EXPLOITATION OF NATIONAL CAPABILITIES PROGRAM—Emerging Requirements, Needs, and Initiatives for Space Operations
—CAPT Robert Huddleston, USN, Director, Navy Technical Exploitation of National Capabilities Program (Navy TENCAP)

1530 HEADQUARTERS MARINE CORPS—Marine Corps Space Systems and Future Developments to Meet Operational Requirements
—BrigGen Richard C. Zilmer, USMC, Director, Strategy and Plans Division, Office of Deputy Commandant for Plans, Policies, and Operations, Headquarters Marine Corps

Closed Session: Committee Members and NRC Staff Only

1700 COMMITTEE DISCUSSION—Summary of Day One
—Dr. Antonio L. Elias, Committee Co-Chair
—ADM William D. Smith, USN (Ret.), Committee Co-Chair

Friday, June 27, 2003

Closed Session: Committee Members and NRC Staff Only

0815 CONVENE—Opening Remarks
—Dr. Antonio L. Elias, Committee Co-Chair
—ADM William D. Smith, USN (Ret.), Committee Co-Chair

Data-Gathering Meeting Not Open to the Public: Classified Discussion (Secret)

0830 NAVAL RESEARCH LABORATORY—Science and Technology Initiatives to Support the Navy's Role in Space
—Mr. Peter G. Wilhelm, Director, Naval Center for Space Technology, Naval Research Laboratory

1000 OPNAV SPACE, INFORMATION WARFARE, COMMAND AND CONTROL DIVISION—Navy Space Requirements, Status of Naval Space Policy
—RDML (Sel) Elizabeth A. Hight, USN, Prospective Deputy for Resources and Requirements; Space, Information Warfare, Command and Control Division; Office of the Chief of Naval Operations, N61R
—Mr. Joseph L. Corcoran, Deputy Head, Space and Communications Branch; Space, Information Warfare, Command

and Control Division; Office of the Chief of Naval Operations, N611B

Closed Session: Committee Members and NRC Staff Only

1230 COMMITTEE DISCUSSION—Continued Discussion of Terms of Reference, Study Plans, Other Issues
—Dr. Antonio L. Elias, Committee Co-Chair
—ADM William D. Smith, USN (Ret.), Committee Co-Chair
—Dr. Charles F. Draper, Associate Director, NSB

Data-Gathering Meeting Not Open to the Public: Classified Discussion (Secret)

1330 ONR—OCEAN, ATMOSPHERE AND SPACE DEPARTMENT—Space Programs and Initiatives at the Office of Naval Research
—Dr. Robert McCoy, Ocean, Atmosphere and Space Department; Office of Naval Research (Code 32)

1430 OPNAV RESOURCES, REQUIREMENTS, AND ASSESSMENTS—Navy Operational Requirements and Programming in Space
—VADM Michael G. Mullen, USN, Deputy Chief of Naval Operations for Resources, Requirements and Assessments, N8

Closed Session: Committee Members and NRC Staff Only

1530 COMMITTEE DISCUSSION—Summary of Day Two
—Dr. Antonio L. Elias, Committee Co-Chair
—ADM William D. Smith, USN (Ret.), Committee Co-Chair

1600 ADJOURN

JULY 28-29, 2003
KECK CENTER OF THE NATIONAL ACADEMIES
WASHINGTON, D.C.

Monday, July 28, 2003

Closed Session: Committee Members and NRC Staff Only

0830 CONVENE—Opening Remarks, Committee Discussion
—Dr. Antonio L. Elias, Committee Co-Chair
—ADM William D. Smith, USN (Ret.), Committee Co-Chair
—Dr. Michael L. Wilson, Program Officer, NSB

Data-Gathering Meeting Not Open to the Public: Classified Discussion (Secret)

0900 OFFICE OF NAVAL INTELLIGENCE—Naval Intelligence, Surveillance, and Reconnaissance Plans and Initiatives and Their Relationship to Space

—RADM Richard B. Porterfield, USN, Director, Office of Naval Intelligence, Office of the Chief of Naval Operations, N2

1030 MARINE CORPS INTELLIGENCE ACTIVITY—Marine Corps Intelligence, Surveillance, and Reconnaissance Plans and Initiatives and Their Relationship to Space
—Mr. Michael Decker, Assistant Director, Intelligence Division, Headquarters, U.S. Marine Corps

Closed Session: Committee Members and NRC Staff Only

1200 COMMITTEE DISCUSSION
Moderators:
—Dr. Antonio L. Elias, Committee Co-Chair
—ADM William D. Smith, USN (Ret.), Committee Co-Chair

Data-Gathering Meeting Not Open to the Public: Classified Discussion (Secret)

1400 MARINE CORPS COMBAT DEVELOPMENT COMMAND—Marine Corps Concepts and Space
—LtCol Gary Russell, USMC, Branch Head, Concepts Branch, Marine Corps Combat Development Command

Closed Session: Committee Members and NRC Staff Only

1530 COMMITTEE DISCUSSION—Plans Ahead, Report Deliberations
Moderators:
—Dr. Antonio L. Elias, Committee Co-Chair
—ADM William D. Smith, USN (Ret.), Committee Co-Chair

1700 END SESSION

Tuesday, July 29, 2003

Closed Session: Committee Members and NRC Staff Only

0830 CONVENE—Opening Remarks, Committee Discussion
—Dr. Antonio L. Elias, Committee Co-Chair
—ADM William D. Smith, USN (Ret.), Committee Co-Chair
—Dr. Michael L. Wilson, Program Officer, NSB

Data-Gathering Meeting Not Open to the Public: Classified Discussion (Secret)

0900 NAVY WARFARE DEVELOPMENT COMMAND—The Role of Space Capabilities in the Sea Trial Experimentation Campaign
—Mr. Wayne I. Perras, Director for Transformation, Navy Warfare Development Command

—CAPT Kevin Morrissey, USN, Director, Maritime Battle Center, Navy Warfare Development Command
—Mr. Roy C. Evans, Jr., Concept-based Experimentation Development, Navy Warfare Development Command

1030 OFFICE OF THE CHIEF OF NAVAL OPERATIONS—Assessment of Naval Capability Gaps and Resulting Needs for Space
—CDR John C. Oberst, USN, Assessment Division, Office of the Chief of Naval Operations, N81

1230 NATIONAL AEROSPACE INITIATIVE—Background, Technology Initiatives
—Mr. Paul Piscopo, National Aerospace Initiative, Special Assistant to the Director of Defense for Research and Engineering

1400 OFFICE OF NAVAL RESEARCH—Naval Research Initiatives and Their Relationship to Space
—RADM Jay M. Cohen, USN, Chief of Naval Research, Office of Naval Research; Director, Test and Evaluation and Technology Requirements, Office of the Chief of Naval Operations, N091

Closed Session: Committee Members and NRC Staff Only

1530 COMMITTEE DISCUSSION—Meeting Summary, Plans Ahead, Report Deliberations
Moderators:
—Dr. Antonio L. Elias, Committee Co-Chair
—ADM William D. Smith, USN (Ret.), Committee Co-Chair

1600 ADJOURN

AUGUST 26, 2003
KECK CENTER OF THE NATIONAL ACADEMIES
WASHINGTON, D.C.

Tuesday, August 26, 2003

Closed Session: Committee Members and NRC Staff Only

0830 CONVENE—Opening Remarks, Committee Discussion
—Dr. Antonio L. Elias, Committee Co-Chair
—ADM William D. Smith, USN (Ret.), Committee Co-Chair
—Dr. Michael L. Wilson, Program Officer, NSB

Data-Gathering Meeting Not Open to the Public: Classified Discussion (Secret)

0900 NAVAL NETWORK AND SPACE OPERATIONS COMMAND—Operation, Coordination, and Development of Naval Space Assets
—RADM John P. Cryer III, USN, Commander, Naval Network and Space Operations Command

1030 NATIONAL OCEANIC AND ATMOSPHERIC ADMINISTRATION—National Environmental Observation Systems and Plans
—Mr. Gary K. Davis, Director, Office of Systems Development, National Oceanic and Atmospheric Administration

1230 OFFICE OF THE OCEANOGRAPHER OF THE NAVY—Naval Oceanography Program and Its Relation to Meteorology and Oceanography (METOC) and Navigation Plans
—CAPT Robert Lawson, USN, Deputy Oceanographer of the Navy, Office of the Chief of Naval Operations, N096B

1530 DEFENSE ADVANCED RESEARCH PROJECTS AGENCY—SPECIAL PROJECTS OFFICE—Programs and Initiatives Related to Advanced Space-Based Technologies
—Dr. Amy Alving, Director, Special Projects Office, Defense Advanced Research Projects Agency

Closed Session: Committee Members and NRC Staff Only

1630 COMMITTEE DISCUSSION
Moderators:
—Dr. Antonio L. Elias, Committee Co-Chair
—ADM William D. Smith, USN (Ret.), Committee Co-Chair

1800 END SESSION

AUGUST 27-28, 2003
NATIONAL RECONNAISSANCE OFFICE
CHANTILLY, VIRGINIA

Wednesday, August 27, 2003

Closed Session: Committee Members and NRC Staff Only

0810 CONVENE—Opening Remarks
—Dr. Antonio L. Elias, Committee Co-Chair
—ADM William D. Smith, USN (Ret.), Committee Co-Chair
—Dr. Charles F. Draper, Acting Director, NSB

Data-Gathering Meeting Not Open to the Public: Classified Discussion (Top Secret+)

0820 INTRODUCTION TO THE NATIONAL RECONNAISSANCE OFFICE (NRO) AND SPAWAR SPACE FIELD ACTIVITY
—Dr. Jack Breedlove, Executive Director, SPAWAR Space Field Activity

0830 NAVAL NETWORK WARFARE COMMAND (NETWARCOM)—Overview of the Role of NETWARCOM in the Support of Space Systems and Activities
—VADM Richard W. Mayo, USN, Commander, Naval Network Warfare Command

0930 NRO SYSTEMS BRIEF

1015 NRO SIGNALS INTELLIGENCE SYSTEMS

1115 NRO COMMUNICATIONS SYSTEMS

1145 TRANSFORMATIONAL COMMUNICATIONS ARCHITECTURE

1300 NRO IMAGERY INTELLIGENCE SYSTEMS

1415 ADVANCED SYSTEMS AND TECHNOLOGY OFFICE

Closed Session: Committee Members and NRC Staff Only

1515 COMMITTEE DISCUSSION—Meeting Summary, Plans Ahead, Report Deliberations
Moderators:
—Dr. Antonio L. Elias, Committee Co-Chair
—ADM William D. Smith, USN (Ret.), Committee Co-Chair

1700 END SESSION

Thursday, August 28, 2003

Data-Gathering Meeting Not Open to the Public: Classified Discussion (Top Secret+)

0800 OPERATIONAL SUPPORT OFFICE—NAVAL TOPICS

0900 JOINT FIRES NETWORK

1015 MARINE CORPS SPACE ISSUES

1045 NAVY TENCAP OFFICE

1200 NAVY-NRO COORDINATION GROUP

Closed Session: Committee Members and NRC Staff Only

1400 COMMITTEE DISCUSSION—Meeting Summary, Plans Ahead, Report Deliberations
Moderators:
—Dr. Antonio L. Elias, Committee Co-Chair
—ADM William D. Smith, USN (Ret.), Committee Co-Chair

1500 ADJOURN

SEPTEMBER 30, 2003
NAVAL RESEARCH LABORATORY
WASHINGTON, D.C.

Tuesday, September 30, 2003

Closed Session: Committee Members and NRC Staff Only

0750 CONVENE—Opening Remarks, Committee Discussion
—Dr. Antonio L. Elias, Committee Co-Chair
—ADM William D. Smith, USN (Ret.), Committee Co-Chair
—Dr. Charles F. Draper, Acting Director, NSB

Data-Gathering Meeting Not Open to the Public: Classified Discussion (Top Secret+)

0800 NAVAL RESEARCH LABORATORY—Naval Center for Space Technology Initiatives; Higher Classification Presentations and Discussion
—Mr. Peter Wilhelm, Director, Naval Center for Space Technology, Naval Research Laboratory

Data-Gathering Meeting Not Open to the Public: Classified Discussion (Secret)

1000 NAVAL RESEARCH LABORATORY—Welcome and Overview of the Naval Center for Space Technology; Programs and Facilities in Support of Naval Needs
—Mr. Peter Wilhelm, Director, Naval Center for Space Technology, Naval Research Laboratory

1300 NAVAL RESEARCH LABORATORY (continued)
—Mr. Peter Wilhelm, Director, Naval Center for Space Technology, Naval Research Laboratory

Closed Session: Committee Members and NRC Staff Only

1600 COMMITTEE DISCUSSION—Plans Ahead, Report Deliberations
Moderators:
—Dr. Antonio L. Elias, Committee Co-Chair
—ADM William D. Smith, USN (Ret.), Committee Co-Chair

1700 END SESSION

OCTOBER 1-2, 2003
KECK CENTER OF THE NATIONAL ACADEMIES
WASHINGTON, D.C.

Wednesday, October 1, 2003

Closed Session: Committee Members and NRC Staff Only

0830 CONVENE—Opening Remarks, Committee Discussion
—Dr. Antonio L. Elias, Committee Co-Chair
—ADM William D. Smith, USN (Ret.), Committee Co-Chair
—Dr. Michael L. Wilson, Program Officer, NSB

Data-Gathering Meeting Not Open to the Public: Classified Discussion (Secret)

0930 SPACE AND NAVAL WARFARE SYSTEMS COMMAND—Space Initiatives in Support of Naval Needs
—RADM Kenneth Slaght, USN, Commander, Space and Naval Warfare Systems Command Control Division, Office of the Chief of Naval Operations, N611

1200 OFFICE OF THE CHIEF OF NAVAL OPERATIONS—INTELLIGENCE, SURVEILLANCE, RECONNAISSANCE REQUIREMENTS (ISR), RESOURCES, AND PROGRAMS DIVISION—Naval Requirements for Space-based Radar and Naval Fires Network
—Mr. Keith Barber, Intelligence, Surveillance, Reconnaissance Requirements (ISR), Resources, and Programs Division, Office of the Chief of Naval Operations, N20

Closed Session: Committee Members and NRC Staff Only

1300 COMMITTEE DISCUSSION—Report Deliberations
Moderators:
—Dr. Antonio L. Elias, Committee Co-Chair
—ADM William D. Smith, USN (Ret.), Committee Co-Chair

Data-Gathering Meeting Not Open to the Public: Classified Discussion (Secret)

1400 OFFICE OF THE CHIEF OF NAVAL OPERATIONS—SPACE, INFORMATION WARFARE, COMMAND AND CONTROL DIVISION—Naval Space Policy Development
—Mr. Joseph L. Corcoran, Deputy Head, Space and Communications Branch; Space, Information Warfare, Command and Control Division, Office of the Chief of Naval Operations, N611B

1515 NAVAL SPACE INITIATIVES AND DEVELOPMENTS
—Dr. Bruce Wald, Arlington Consulting Associates
—Dr. Gary Federici, Center for Naval Analyses

Closed Session: Committee Members and NRC Staff Only

1615 COMMITTEE DISCUSSION—Meeting Summary, Plans Ahead, Report Deliberations
Moderators:
—Dr. Antonio L. Elias, Committee Co-Chair
—ADM William D. Smith, USN (Ret.), Committee Co-Chair

1730 ADJOURN

Thursday, October 2, 2003

Closed Session: Committee Members and NRC Staff Only

0830 CONVENE—Opening Remarks, Committee Discussion
—Dr. Antonio L. Elias, Committee Co-Chair
—ADM William D. Smith, USN (Ret.), Committee Co-Chair
—Dr. Michael L. Wilson, Program Officer, NSB

Data-Gathering Meeting Not Open to the Public: Classified Discussion (Secret)

0930 NAVY TECHNICAL EXPLOITATION OF NATIONAL CAPABILITIES (TENCAP) PROGRAM—Requirements in Support of Navy
—Mr. Frederick J. Glaeser, Navy Technical Exploitation of National Capabilities Program (Navy TENCAP)

1030 BOEING CONNEXION—Ku Band Satellite Communications Capabilities in Support of Naval Needs
—Mr. Gerald S. Hopp, Government Sales Director, The Boeing Company

1200 CENTER FOR NAVAL ANALYSIS—Understanding Afloat Network Performance and Usage
—Mr. Dennis Shea, Research Team Leader, Center for Naval Analyses
—Dr. John Bentrup, Research Analyst, Center for Naval Analyses

1300 SPACE-BASED RADAR—Air Force Development of Space-Based Radar in Support of Department of Defense Needs
—Brig Gen John T. (Tom) Sheridan, USAF, Director of Requirements, Air Force Space Command

Closed Session: Committee Members and NRC Staff Only

1400 COMMITTEE DISCUSSION—Meeting Summary, Plans Ahead, Report Deliberations

Moderators:

—Dr. Antonio L. Elias, Committee Co-Chair

—ADM William D. Smith, USN (Ret.), Committee Co-Chair

1500 ADJOURN

OCTOBER 28-29, 2003
KECK CENTER OF THE NATIONAL ACADEMIES
WASHINGTON, D.C.

Tuesday, October 28, 2003

Closed Session: Committee Members and NRC Staff Only

0745 CONVENE—Opening Remarks, Committee Discussion

—Dr. Antonio L. Elias, Committee Co-Chair

—ADM William D. Smith, USN (Ret.), Committee Co-Chair

—Dr. Michael L. Wilson, Program Officer, NSB

Data-Gathering Session Not Open to the Public: Classified Discussion (Secret)

0800 JOINT CHIEFS OF STAFF—Joint Requirements and Programming for Space Systems and Space Operations

—Lt Col Michael LaPointe, USAF, Weapons Systems Program Evaluation; Force Structure, Resources, and Assessments; U.S. Joint Chiefs of Staff, J8

—Maj Kenneth Myers, USAF, Space Systems Division; Command, Control, Communications, and Computing Directorate; U.S. Joint Chiefs of Staff, J6

0900 NAVAL SPECIAL WARFARE COMMAND—Special Warfare Operational Needs and Initiatives Aimed at Exploitation of Space Systems

—CDR Wanda Janus, USN, Chief Information Officer, Naval Special Warfare Command

1030 NAVSTAR GLOBAL POSITIONING SYSTEM JOINT PROGRAM OFFICE—GPS III—Development, Programming, and Operational Needs for the Next Generation GPS System

—Col Roger E. Robb, USAF, Deputy System Program Manager, Navstar Global Positioning System Joint Program Office

Closed Session: Committee Members and NRC Staff Only

1130 COMMITTEE DISCUSSION—Report Deliberations, Chapter Presentations and Discussion
Moderators:
—Dr. Antonio L. Elias, Committee Co-Chair
—ADM William D. Smith, USN (Ret.), Committee Co-Chair

Data-Gathering Session Not Open to the Public: Classified Discussion (Secret)

1430 UNDER SECRETARY OF DEFENSE FOR INTELLIGENCE—Department of Defense Intelligence, Surveillance, and Reconnaissance Plans and Initiatives and Their Relationship to Space Systems
—Mr. Kevin Meiners, Director, Intelligence, Strategies, Assessments, and Technology (ISAT); Office of the Deputy Under Secretary of Defense (Intelligence) for Intelligence and Warfighting Support

Closed Session: Committee Members and NRC Staff Only

1600 COMMITTEE DISCUSSION—Plans Ahead; Continue Report Deliberations, Chapter Presentations and Discussion
Moderators:
—Dr. Antonio L. Elias, Committee Co-Chair
—ADM William D. Smith, USN (Ret.), Committee Co-Chair

1730 END SESSION

Wednesday, October 29, 2003

Closed Session: Committee Members and NRC Staff Only

0830 CONVENE—Opening Remarks, Committee Discussion
—Dr. Antonio L. Elias, Committee Co-Chair
—ADM William D. Smith, USN (Ret.), Committee Co-Chair
—Dr. Michael L. Wilson, Program Officer, NSB

Data-Gathering Session Not Open to the Public: Classified Discussion (Secret)

0900 TRANSFORMATIONAL COMMUNICATIONS ARCHITECTURE (TCA)—Naval Involvement in TCA; Tactical and Strategic Needs
—RADM Rand H. Fisher, USN, Commander, SPAWAR Space Field Activity; Director, Naval Space Technology Programs

Closed Session: Committee Members and NRC Staff Only

1030 COMMITTEE DISCUSSION—Report Deliberations, Chapter Presentations, and Discussion

Moderators:

—Dr. Antonio L. Elias, Committee Co-Chair

—ADM William D. Smith, USN (Ret.), Committee Co-Chair

Data-Gathering Session Not Open to the Public: Classified Discussion (Secret)

1230 DISCUSSION ON DEPARTMENT OF NAVY FUTURE SPACE SYSTEMS—Concept Development, Experimentation, Requirements Generation, and Science and Technology Needs

Participants:

—Mr. Keith Barber, Intelligence, Surveillance, Reconnaissance Requirements (ISR), Resources, and Programs Division, Office of the Chief of Naval Operations, N20

—CDR Randy Bell, USN, Space-Based Radar Program Office, Space and Missile Center

—RADM Jay M. Cohen, USN, Chief of Naval Research, Office of Naval Research; Director, Test and Evaluation and Technology Requirements, Office of the Chief of Naval Operations, N091

—RADM John M. Kelly, USN, Commander, Navy Warfare Development Command

—RADM Rand Fisher, USN, Commander, SPAWAR Space Field Activity; Director, Naval Space Technology Programs

—CDR Maria Lyles, USN, Director, Navy Technical Exploitation of National Capabilities Program (Navy TENCAP)

—CAPT David Markham, USN, Chief of Staff for Advanced Systems and Technology, National Reconnaissance Office

—Mr. Wayne I. Perras, Director for Transformation, Navy Warfare Development Command

—RADM Steven J. Tomaszeski, USN, Oceanographer of the Navy, Office of the Chief of Naval Operations, N096

—Mr. Peter Wilhelm, Director, Naval Center for Space Technology, Naval Research Laboratory

—RADM Thomas E. Zelibor, USN, Director, Space, Information Warfare, Command and Control Division, Office of the Deputy Chief of Naval Operations for Warfare Requirements and Programs, N61

—BrigGen Richard C. Zilmer, USMC, Director, Strategy and Plans Division, Office of the Deputy Commandant for Plans, Policies and Operations

1400 END SESSION

Panel Discussion Session Not Open to the Public: Classified Discussion (Top Secret+)

1415 DEPARTMENT OF NAVY FUTURE SPACE SYSTEMS—Focus on Space-Based Radar as a Test Case for Navy Input
—Continuation with above participants

Closed Session: Committee Members and NRC Staff Only

1530 COMMITTEE DISCUSSION—Meeting Summary, Plans Ahead, Report Deliberations
Moderators:
—Dr. Antonio L. Elias, Committee Co-Chair
—ADM William D. Smith, USN (Ret.), Committee Co-Chair

1700 ADJOURN

NOVEMBER 14, 2003
KECK CENTER OF THE NATIONAL ACADEMIES
WASHINGTON, D.C.

Friday, November 14, 2003

Closed Session: Committee Member and NRC Staff Only

0815 CONVENE
—Dr. Joseph Reagan, Committee Member
—Dr. Michael Wilson, Study Director

Data-Gathering Session Not Open to the Public: Classified Discussion (Secret)

0830 SPACE-BASED RADAR—Development, Programming, and Operational Needs for the Space-Based Radar System
—LtCol Ronald A. Grundman, USAF, Deputy Program Director, Space-Based Radar Joint Program Office

1000 U.S. NAVAL POSTGRADUATE SCHOOL—The Space Sciences Program and Its Relation to the Naval Space Cadre
—Dr. Rudolf Panholzer, Dean, Graduate School of Engineering and Applied Sciences, and Chairman, Space Systems Academic Group, U.S. Naval Postgraduate School

1230 MOBILE USER OBJECTIVE SYSTEM—Development, Programming, and Operational Needs for the Mobile User Objective System
—Mr. Robert Tarleton, Program Manager, Communications Satellite Program Office, Space and Naval Warfare Systems Command, PMW-146

—Ms. Maureen Jackson, Washington Liaison Officer, Communications Satellite Program Office, Space and Naval Warfare Systems Command, PMW-146

1400 UNDERSECRETARY OF DEFENSE FOR INTELLIGENCE—Horizontal Integration Initiatives and Their Relationship to Developing Space Programs
—Col Kevin McLaughlin, USAF, Director for Planning and Preparation, Office of the Deputy Undersecretary of Defense for Preparation and Warning, Office of the Under Secretary of Defense for Intelligence

1530 ADJOURN

DECEMBER 1-5, 2003
ARNOLD AND MABEL BECKMAN CENTER
IRVINE, CALIFORNIA

Monday, December 1, 2003

Closed Session: Committee Members and NRC Staff Only

0830 CONVENE—Welcome, Administrative Issues, Meeting Schedule
—Antonio L. Elias, Committee Co-Chair
—William D. Smith, Committee Co-Chair
—Michael Wilson, NSB Program Officer

0900 CHAPTER 1 BRIEF—Introduction
—Antonio L. Elias, Committee Co-Chair
—William D. Smith, Committee Co-Chair

1000 CHAPTER 2 BRIEF—Operational Concepts and Needs
—Alan Berman, Chapter 2, Lead Author

1100 CHAPTER 3 BRIEF—Roles and Responsibilities
—Gene Porter, Chapter 3, Lead Author

1300 CHAPTER 4 BRIEF—Operational Functions Provided from Space: ISR
—Dell Williams, Chapter 4, ISR Lead Author

1330 CHAPTER 4 BRIEF—Operational Functions Provided from Space: METOC
—Ann Berman, Chapter 4, METOC Lead Author

1400 CHAPTER 4 BRIEF—Operational Functions Provided from Space: Missile Defense
—Alan Berman, Chapter 4, Missile Defense Lead Author

1430 CHAPTER 4 BRIEF—Operational Functions Provided from Space: Communications
—George Nossaman, Chapter 4, Communications Lead Author

1500 CHAPTER 4 BRIEF—Operational Functions Provided from Space: Position, Navigation, Timing
—Greg Tornatore, Chapter 4, Position, Navigation, Timing Lead Author

1530 CHAPTER 4 BRIEF—Operational Functions Provided from Space: Space Control
—Don Hard, Chapter 4, Space Control Lead Author

1600 CHAPTER 5 BRIEF—Fulfilling Naval Forces Space Needs
—David Whelan, Chapter 5, Lead Author

1700 WRAP-UP, SCHEDULE FOR DECEMBER 2

1900 END SESSION

December 2-5, 2003

Closed Session: Committee Members and NRC Staff Only

0830 CONVENE—Plans for the Day
—Antonio L. Elias, Committee Co-Chair
—William D. Smith, Committee Co-Chair
—Michael Wilson, NSB Program Officer

0900 REPORT DISCUSSION AND DRAFTING

1700 END SESSION

G

Acronyms and Abbreviations

ABL	Airborne Laser
A/C	aircraft
ACTD	Advanced Concept Technology Demonstration
ACTS	Advanced Communications Technology Satellite
ADNS	Advanced Digital Network System
Adv Polar	Advanced Polar Satellite
AEHF	Advanced Extremely High Frequency (satellite)
AISR	airborne intelligence, surveillance, and reconnaissance
AMTI	airborne moving target indication
APL	Applied Physics Laboratory (Johns Hopkins University)
ARPA	Advanced Research Projects Agency
ASCMD	antiship cruise missile defense
ASD(NII)	Assistant Secretary of Defense for Networks and Information Integration
ASW	antisubmarine warfare
ATPA	active transmit phased array
AWACS	Airborne Warning and Control System
AWS	Advanced Wideband System
BDA	battle damage assessment
BFT	blue force tracking
BLOS	beyond line of sight
BMD	ballistic missile defense

C4ISR	command, control, communications, computers, intelligence, surveillance, and reconnaissance
C/A-code	coarse acquisition code (GPS signal)
CBRNE	chemical, biological, radiological, nuclear, and environmental
CEC	Cooperative Engagement Capability
CFFC	Commander, Fleet Forces Command
CIA	Central Intelligence Agency
CINC	Commander in Chief
CMIS	Conical Scanning Microwave Imager/Sounder
CNO	Chief of Naval Operations
CNR	Chief of Naval Research
COMSAT	communications satellite
COTP	common operational and tactical picture
CVN	aircraft carrier, nuclear-powered
CWSP	Commercial Wideband Satellite Program
DAHS	Dual Antenna Handover System
DAMA	demand assigned multiple access
DARPA	Defense Advanced Research Projects Agency
DBS	Digital Broadcast Service
DGPS	Differential Global Positioning System
DISA	Defense Information Systems Agency
D/L	down link
DLA	Defense Logistics Agency
DMSP	Defense Meteorological Satellite Program
DOD	Department of Defense
DOE	Department of Energy
DOTMLPF	doctrine, organization, training, material, leadership development, personnel, and facilities
DSCS	Defense Satellite Communications System
DSP	Defense Support Program
DSTS	Defense Satellite Transmission Service
DTS	Direct Transmission System
EHF	extremely high frequency
ELINT	electronic intelligence
EMI	electromagnetic interference
EMW	Expeditionary Maneuver Warfare
EO/IR	electro-optic/infrared
Eutelsat	European telecommunications satellite
FAB-T	family of advanced beyond-line-of-sight terminals
FBP	Fleet Broadcast Processor

FEP	Fleet EHF Package
FIA	Future Imagery Architecture
FLTSAT	Fleet Satellite
FNC	Future Naval Capabilities
FNMOC	Fleet Numerical Meteorology and Oceanography Center
FOT	Fleet Operational Terminal
FSS	Fixed Satellite Service
GBS	Global Broadcast Service
GDA	gimbaled dish antenna
Geosat	Geodetic Satellite
GFO	Geosat Follow-on
GIFTS	Geosynchronous Imaging Fourier Transform Spectrometer
GIG	Global Information Grid
GIG-BE	Global Information Grid Bandwidth Expansion
GLONASS	Global Orbiting Navigation Satellite System
GMS	Geostationary Meteorological Satellite
GMT	ground multiband terminal
GMTI	ground moving target indication
GOES	Geostationary Operational Environmental Satellite
GPS	Global Positioning System
GRAB	Galactic Radiation and Background (satellite)
HAIPE	High Assurance Internet Protocol Encryptor
HEMP	high-altitude electromagnetic pulse
HF	high frequency
HMMWV	high-mobility multipurpose wheeled vehicle
IGY	International Geophysical Year
IMINT	imagery intelligence
Inmarsat	International Maritime Satellite (organization)
INSAT	Indian National Satellite
Intelsat	intelligence satellite
IOMI	Indian Ocean METOC Imager
IP	Internet Protocol
ISR	intelligence, surveillance, and reconnaissance
JCIDS	Joint Capabilities Integration and Development System
JPO	Joint Program Office
JROC	Joint Requirements Oversight Council
JTRS	Joint Tactical Radio System
LC	laser crosslink

LCC	command ship
LCS	littoral combat ship
LDR	low data rate
LIDAR	light detection and ranging
(L)MSS	(Land) Mobile Satellite Service
LMST	lightweight multiband satellite terminal
LORAN	Long Range Navigation (system)
LPD	low probability of detection
LPI	low probability of intercept
Mbp/s	megabits per second
M-code	military only (GPS signal)
MDA	Missile Defense Agency
MDR	medium data rate
Meteosat	Meteorological Satellite
METOC	meteorology and oceanography
MILSATCOM	Military Satellite Communications
MILSTAR	Military Strategic, Tactical, and Relay (satellite)
MIST	Multiband/multimode Integrated Satellite Terminal
MODIS	Moderate Resolution Imaging Spectro-radiometer
MSC	Military Sealift Command
MTI	moving target indicator
MUOS	Mobile User Objective System
NAD	Navy Area Defense
NASA	National Aeronautics and Space Administration
NAVAIR	Naval Air Systems Command
NAVO	Naval Oceanographic Office
NAVSEA	Naval Sea Systems Command
NAVSPASUR	Naval Space Surveillance (system)
NAVSTAR	Navigation Satellite Timing and Ranging
NCST	Naval Center for Space Technology
NDS	National Development Satellite
NEMO	Naval EarthMap Observer
NETWARCOM	Naval Network Warfare Command
NGA	National Geospatial-Intelligence Agency
NII	Networks and Information Integration
NMOC	Naval Meteorology and Oceanography Command
NNSOC	Naval Network and Space Operations Command
NOAA	National Oceanic and Atmospheric Administration
NOC	Naval Operating Concept for Joint Operations
NPOESS	National Polar-orbiting Operational Environmental Satellite System

NPS	Naval Postgraduate School
NRC	National Research Council
NRL	Naval Research Laboratory
NRO	National Reconnaissance Office
NSA	National Security Agency
NSS	National Security Space
NSWC	Naval Surface Warfare Center
NTM	National Technical Means
NTS	Navigation Technology Satellite
NTW	Navy Theater Wide
NWDC	Navy Warfare Development Command
OCMD	overland cruise missile defense
OCS	operational control segment
OMFTS	Operational Maneuver From the Sea
ONR	Office of Naval Research
OPARS	Optimum Aircraft Routing System
OPNAV	Office of the Chief of Naval Operations
OPnet	Operational Network
ORBCOMM	satellite data communications system
ORCA	Optical Communications Relay Architecture
OSD	Office of the Secretary of Defense
OTH	over the horizon
OTSR	optimum track ship routing
Panamsat	Pan American satellite
P-code	precision code (GPS signal)
PCS	Personal Communications Service
PGM	precision-guided munition
PNT	position, navigation, and timing
POES	Polar Operational Environmental Satellite
PPS	precise positioning service
PRN	pseudo-random noise
R&D	research and development
RCS	radar cross-section
RF	radio frequency
RFI	radio frequency interference
RMP	Radar Modernization Program
S&T	science and technology
SA	selective availability
SAIL	Special Annotated Imagery-Littoral

SAR	synthetic aperture radar
SATCOM	satellite communications
SBIRS	Space-Based Infrared System
SBIRS-H	Space-Based Infrared System-High
SBIRS-L	Space-Based Infrared System-Low
SBR	Space Based Radar
SDARS	Satellite Digital Audio Service
SECNAV	Secretary of the Navy
SEP	spherical error probable
SGLS	Satellite-Ground Link System
SHF	super high frequency
SIGINT	signals intelligence
SLEP	Service Life Extension Program
SM	Standard Missile
SMEI	Solar Mass Ejection Imager
SOF	special operations forces
SOLRAD	Solar Radiation (satellite)
SPAWAR	Space and Naval Warfare Systems Command
SPS	standard positioning service
SSM/I	Special Sensing Microwave/Imager
SSN	attack submarine
SSPO	Strategic Systems Program Office
STOICS	Special Tactical Oceanographic Information Charts
STOM	Ship-to-Objective Maneuver
STP	Space Test Program
SYSCOM	systems command
TacSat	Tactical Microsatellite
TARPS	tactical airborne reconnaissance pod system
TBD	to be determined
TBM	theater ballistic missile
TBMD	theater and ballistic missile defense
TC	Transformational Communications
TCA	Transformational Communications Architecture
TDRSS	Tracking and Data Relay Satellite System
TENCAP	Technical Exploitation of National Capabilities Program
TIROS	Television and Infrared Observation Satellite
TPED	tasking, processing, exploiting, and disseminating
TPPU	tasking, posting, processing, and using
TRAnet	Transit Network
TSAT	Transformational Communications Satellite
TV-DTS	television-digital telecommunications system

UAV	unmanned aerial vehicle
UFO	UHF Follow-on
UHF	ultrahigh frequency
USAF	U.S. Air Force
USB	Unified S-Band
USN	U.S. Navy
USNO	U.S. Naval Observatory
VHF	very high frequency
WBS	Wideband System
WGS	Wideband Gapfiller System
WSMR	White Sands Missile Range
XDR	expanded data rate
Y-code	encrypted P-code (GPS signal)